335TH

ASSAULT

WHAT WE DID
AFTER THE
VIETNAM WAR

HELICOPTER

COMPANY

VANCE
GAMMONS
& DOMINIC
FINO

Deeds Publishing | Atlanta

Published by Deeds Publishing in Athens, GA
www.deedspublishing.com

Printed in The United States of America

Cover design by Mark Babcock

Memorial photos by Jenna Adaway Photography, www.adawayphotography.com

Library of Congress Cataloging-in-Publications data is available upon request.

ISBN 978-1-947309-38-8

Books are available in quantity for promotional or premium use. For information, email info@deedspublishing.com.

First Edition, 2018

10 9 8 7 6 5 4 3 2 1

DEDICATION

The 335th came into existence September 1966 with the deactivation of Co. A 82nd Aviation Battalion. Co. A arrived in Vietnam April 1965 to support the 173rd Airborne Brigade in III Corps. They continued the mission until the 335th Aviation Co. was formed at Bien Hoa. In early 1967, the 335th was renamed the 335th Assault Helicopter Company.

In May 1967, the 335th moved to Pleiku in II Corps with the 173rd Airborne. In August 1967, they relocated to Phu Hiep (also in II Corps). From this base they supported the 173rd Airborne in its desperate fighting against the NVA and VC around Dak To that November. In December 1968, the 335th relocated to III Corps at Bear Cat. In August 1970, they moved to Dong Tam in IV Corps (the Delta). In November 1971, they stood down and returned to the US.

This work is dedicated to those who served with or were attached to the 335th Assault Helicopter Company.

FOREWORD

When I met Vance Gammons in Vietnam, he was the Company Commander of the 335th Assault Helicopter Company known as the Cowboys. At that time, none of us knew what would become of our lives if we were lucky enough to survive our time in country. At the 2017 Cowboy Reunion, Vance suggested the Cowboys document what they did after leaving Vietnam. A sociological exercise in what might be called, what did you do after the war? This collection of biographies is their story after Vietnam and will inspire those that read it to always push forward and never forget those with whom they served. It is truly a work 50-years in the making.

This journey began in the early 1990s when one Cowboy contacted another Cowboy and they discussed what they had been up to and asked the big question: Whatever happened to so and so? That impromptu process led to a mission to find as many Cowboys as possible for a reunion. For the next couple of years, the search continued, and the first Cowboy Reunion occurred in 1993. The search for Cowboys is nonstop, and a reunion has occurred every two years since.

This collection of biographies, written in their own words, is very compelling. It describes what they accomplished, their ups and downs, and their continued support for our nation. In short, it is a collection of not so famous, ordinary men who survived Vietnam and went on to lead meaningful and productive lives.

Thank you, Vance Gammons, for keeping the spirit alive.

—Dominic Fino

PREFACE

At the 2017 Reunion, former Cowboy Company Commander Vance Gammons planted the seed for this project. Vance thought it would be interesting to read the stories of the Cowboys after they left Vietnam and/or the military. *We certainly know what each of us did while in Vietnam and we talk about life as it was back then. However, few of us know many details of what transpired after Vietnam.*

As usual, Vance made some compelling arguments as to why such a project should be initiated.

ACKNOWLEDGEMENTS

I am grateful for constructive comments and suggestions from Deeds Publishing, and my friend and fellow Vietnam veteran Bob Babcock.

Many have volunteered their time in keeping the Cowboy spirit alive. Here are a few:

- Doug Wilson and "Super" Susan Posey — Buddy Finders dedicated to locating as many Cowboys as possible. Over 700 strong.
- Andy Hooker — Reunion and Special Event Coordinator.
- Russ and Darlene Stibbe — reunion Food and Beverage Coordinators.
- Jack Hunnicutt, Jim Stein and Tom Gould — Coordinators of the Cowboy Memorial Project, which now resides at Veterans Park, Fort Rucker Alabama.
- Dominic Fino — Quarterly Newsletter Editor and HMFIC.

I also wish to thank all the Cowboys that submitted content to this publication. Your contribution and cooperation made this project possible.

Thank you very much, everyone!

— *Vance Gammons*

CONTENTS

INTRODUCTION

The following pages are a snapshot of what the Cowboys did after Vietnam. Their life story in their own words. Life did not end after leaving Vietnam or the US Army. They went on to lead productive lives and contribute to society, along with all those that returned home to a nation that was less than supporting or grateful.

This is a testament to how these men contributed, endured, and overcame many obstacles to merge back into society after their service was complete.

The Cowboys in Vietnam is indeed one of our most telling stories. It shows adaptability, it shows courage and comraderies, it epitomizes in so many ways all that any military organization could wish to be. This is your story.

I have read all the stories that follow, and I hope you will treasure and put them in that box or part of the attic where we hope our children and grandchildren may one day discover and read out of curious interest and be proud of us. They are stories of brave men who did their duty, of families who endured the long months at home wondering about their loved ones, their safety, and whether they would come home. We see the vast landscape of life after the service with experiences spanning from the clergy to education and volunteerism, and the careers for those who continued their Army career. A common thread permeates — pride of duty, pride of leaving your mark, and pride in being part of the Cowboys.

The Cowboys created a history that is unlikely to be repeated—they went where needed, regardless of environment. In addition, the Cowboys completed every mission and never missed a takeoff.

How Vietnam veterans coming home were at times treated is a part of American history that may be described as not our nation's best moment. Many, if not all of you, would have felt this. We cannot erase this past, nor can we erase the tough years some have endured since their service. But in these pages, I see the resilience of the many, the pride that comes with a sense of belonging to something special, and the never removed desire to serve our country.

—Dominic Fino

ALAIMO, BRUCE T.

335th AHC Cowboys from April 1969 – April 1970
Mustang 23 (aka Midnight Cowboy / Warthog)
Vietnam

Assigned to the 2nd platoon Mustangs upon arrival in April 1969. I flew missions for almost two months before I took over Supply Officer duties, replacing Ed Eget in June 1969 prior to the change of command from Maj. Howard Styles to Maj. Vance Gammons. Being Supply Officer along with other additional assignments (i.e. Entertainment Officer, a very tough job as you can imagine) greatly affected my flying time until I passed on the supply officer job to Fred Miller prior to the end of 1969. I still managed to get over 1,000 flying hrs. in country and was fortunate to have made it through with the crews I flew with in one piece! I waited eight months for my first R&R and was fortunate to get Australia. Fortuitously, it was during the Christmas Holidays and during my stay I met my future wife, Jan, on the world-famous Bondi Beach. Having just graduated from college, she was travelling to London in May to work. I returned to the US in April and Jan stopped to visit me and meet my family in Buffalo, before my next assignment. She cancelled her trip and followed me to Fort Rucker and we were married in October 1970.

Military Experience post Vietnam / back to civilian life

At Fort Rucker, I was assigned as a Basic Instrument Instructor and then a stint with the Department of Tactics as a flight instructor until leaving the service in October 1971. We moved to Florida where

3

I completed university B.S. Degrees in Management and Accounting from Central Florida University. I stayed in the Army reserves until leaving the US Army. I did not experience the anti-war sentiment that was present at this time while I was living in Florida and attending college as a mature age student after discharge. There was a large ex-military contingent with young families there. In Australia, there was some resentment for some years before the Vietnam vets were properly respected for their service to country. I lost many brothers in arms during this war and believe they fought for the freedom that the western world enjoys. The coming generations are not being taught that freedom has come at a price and take what was hard won for granted. I cherish my time in the military, the brotherhood of men I fought with, and the life's experiences and training that has helped me to succeed through both good and bad times.

Moving to Australia

In October 1975, after graduating, Jan and I moved to Sydney, Australia with our 2-year old daughter, Nicole. We had another girl, Natalie shortly after and raised our family here to this day. I became an Australian citizen in the late 1990s after the USA allowed Americans to hold dual citizenship. Fortunately, or unfortunately, I cannot run for a political office here holding two citizenships!

Life in Australia

Living in Australia has been a great joy for me in family, work, and play. Having a love of sports and outdoor activities and living in such a temperate climate, this country was a perfect fit. I played competitive basketball early on and then baseball, tennis, and golf year-round to this very day.

When I first arrived and still having family living in the USA, it was expensive to call the parents, let alone travel to the USA with a small family. Fortunately, as my business prospered, we were able to travel to the US regularly for family visits and holidays. On one of our family ski trips in the 1990s to Colorado, we fell in love with the Vail Valley and purchased a second home there. We now enjoy the best of both hemispheres.

The military brotherhood of Australians who served in Vietnam is very strong and supportive. The Australian public is very appreciative of those who serve and have served in the armed forces. I have been welcomed in by the 135th AHC "EMU's" and recently attended their 50-year reunion in October 2017. I have been involved with an Army Officer golf club of serving and retired officers in Sydney for over 20 years and the camaraderie that is shared among them is enduring.

Work Career

I started my business career in Australia as a management consultant for a Sydney advertising agency wanting to start a sign making business. My employer ran into financial trouble soon after it was up and running and I ended up inheriting the business and its assets in lieu of fees owed. My business career in Australia thus started in the sign making industry.

I grew the business over the first 12 years, expanding into Neon signs, screen printing, plastic and metal fabrication before moving into custom corporate displays. In the late 1980s, I expanded the business, purchasing a building company doing retail and commercial fit outs. My workforce grew to 65 skilled tradesmen and admin staff by the early 1990s.

The expansion was profitable, but selling and managing the business units was a never-ending job with a lot of stress. Adding to these long hours, union problems and an economic downturn left me worn out and with a bad taste for operating a multi-faceted business in such a small market. I restructured, utilizing the knowledge gained, and contacts acquired and focused on my managerial and organizational skills incorporating design and project management while using contract labor for construction as needed.

The change provided me with a new lease on life and a rewarding career having the freedom to better manage my work and family life. My work has taken me to all parts of Australia and New Zealand and internationally to China, Japan, and Europe.

2018

Married for 47 years to the beautiful girl I met on Bondi beach, father of two beautiful daughters and four grandchildren. I am thankful for my tour in Vietnam and my fateful R&R in Sydney in 1969. My military experience has made a positive impact on my life which I will never regret. I was proud to serve, and I honor all those who served with me. I am semi-retired now, enjoying a balance of work, family, and travelling between Australia, the USA, and Europe.

ALLUE, EUGENE (GENE) WARREN

Cowboy 3 / Mustang 25

I got back from Vietnam in November of 1969. I started with the Cowboys in November 1968 at Phu Heip and my tour ended with the Cowboys at Bear Cat.

After my tour, I spent four years of active duty at Davison Army Airfield, Fort Belvoir. It was the greatest assignment an Aviator could ask for. There was no Army stuff, just flying every day; flying VIPs around Virginia, Maryland, and Pennsylvania. While there I was able to get a transition in a VH34 and an OH 58. You could schedule the OH 58 and go fly by yourself wherever you wanted to go. I also received training and got an Instrument ticket. I developed a love for Instrument flying, which was a blessing the rest of my flying days.

Davison was a great assignment, but in 1973, I was separated from the Army and I went into real estate, working for a company called Mount Vernon Realty in Virginia. I did fairly well and was able to put enough money together and I started a custom home construction company. I built and sold about 100 custom homes over the next 15 years.

In 1980, I bought a plane and so that I'd have a place to fly the airplane, I joined the Virginia National Guard down at Richmond. Former Cowboy Dennis Staples was in the unit. I was in the Virginia guard until 1990, at which time I took a promotion and transferred to the Maryland National Guard. While in the Maryland guard, I was offered a full-time position as a test pilot. Given this opportunity, I had always said I wanted to fly for a living so I closed the construction company and went to work full-time as a test pilot. I did that until 1994 and then left them and went into the Army Reserves. I stayed for three years and retired as a Lieutenant Colonel.

In the 1980s, while I owned my plane, I flew all over the country. I flew down all through the Bahamas, including the southernmost island of the Bahamas, San Salvador. My family and I flew from Virginia to Vermont to South Dakota, then Denver, Colorado to Brownsville, Texas to Miami, Florida and back to Virginia, all in one trip. This trip took about two months. We just toured the United States, the only regret I have is that I flew at an altitude of about 12,000 feet. For a more exciting time, I should have made that whole flight at 500 feet.

I did have an opportunity to go to Seattle, Washington and pick up an OH 58 while I was in the Maryland Guard. Another pilot and I flew that OH-58 from Seattle, Washington back to Bel Air, Maryland and we never went over 500 feet AGL. It was the greatest flight ever.

In 1994 when I left full-time employment at the National Guard, I went to work for Prosperity Mortgage as a loan officer. I worked there from 1994 until 2011 and retired as an area manager. Since retiring, my back has caused me great problems so I don't do much and I am pretty inactive. I am a member of NRA and VHPA.

As for my military experience, what it did to me that year in Vietnam was the most satisfying year of my life. The friends I met, the people I have thought about for all these years; the experiences I had while we were flying; I remember it was fun. We had great camaraderie, no fights, no arguments, and people just getting along, working together for one goal. The worst memory I have is the early morning when we lost two aircraft at Blackhorse. I was in that flight; it haunts me to this day; the mistakes we made as a group that day. I wish we could redo that morning.

As far as the military goes, I have great respect for the men and women in the military, I always have. However, it's the government

(The Swamp) that I notice is corrupt. I became especially aware in 2012 when we pulled out-of Iraq. Then I thought, *"Wow, we did the same thing in Vietnam."* The sacrifices we made, the lives lost, and because of politics we just picked up and left. We were actually there to help them be free from communism. We did the same thing in Iraq (freedom from a dictator), but we left that country in shambles because our politicians decided to pull out and left those people at the mercy of gangsters, murderers, and worse. So anytime we commit our military, we should be committing it for one reason and one reason only; because it benefits the United States and we are going to stay in it until it's over and it's done.

The girl I married four months before I went to Vietnam, my high school sweetheart, 50 great years later, I'm still married to her (chained). We have two children, and five grandchildren. When I think about Vietnam, I think about how my life has benefited from that experience. I was an operations officer in a company of Great Men who did a fantastic job and today I still think it was the highlight of my life. I often think about Major Stiles and Captain Hernandez and how they entrusted me with the job of being the Operations officer. That one job has shaped my entire life. I have used the experiences and knowledge learned from that year to make my life a success.

AUKERMAN, LARRY A.

I did not go to back to school of any kind.

After my military discharge, I went to work in a magazine factory in Dayton, Ohio for 20 years.

After the shutdowns in the 1980s, I went to work at a Speed shop for the next 20 years.

I enjoyed working on race cars and going to Sprint car races in the tristate area.

I married in 1978 and had two sons and a daughter. Now I have three granddaughters.

I did not want my children to enter the military.

I do not like or attend organizations.

I am a sports fan, watching all my children participate. Now I watch sports every night on cable. I also enjoy classic cars and attend many car shows.

Since retirement, we spend a couple months every winter around Naples, FL.

I have used the VA medical facilities for all my health issues for the past 20 years and I am a cancer survivor.

AVENI, ROGER ALAN

After Vietnam, I was stationed at Fort Wolters as an instructor pilot for about one year and in-flight evaluation for somewhat less than a year. I left the service 26 May 1969 and moved from Weatherford, TX to Austin, TX. I moved to Austin to attend the University of Texas where I received a Bachelor of Science degree in electrical engineering (BSEE) in 1972.

After graduation, I worked for a small company in Austin for a couple of years until the company's product (we built a product that tested the beer level in beer cans as the cans moved along the assembly line) was sold to a larger company.

I moved back to New Hampshire (my home state) in early 1974 where I found work at a defense contractor called Sanders Associates.

From 1974 to 1982, I worked for several companies as an electrical engineer/software developer including Sanders Associates, Wang, and Honeywell.

In early 1982, I started selling real estate part time and found I was making twice as much money and having fun doing it. In 1982, I bought a Century 21 franchise and opened a real estate office near my home town. The real estate market was humming along in the early 1980s and I bought two more Century 21 franchises in neighboring towns.

In 1985, I opened a day care center. How did I get into the day-care business, you ask? A real estate customer was going to open a day care business in a property I owned. The customer backed out of the lease and I decided to open the day care business myself. My daugh-

11

ter Tess was born in early 1985 and she attended the day care through kindergarten.

In the later 1980s, I was involved in land development, two small subdivisions, and a couple of spec houses. The real estate market was not much fun (not profitable) in the early 1990s and I closed two of my offices and sold the remaining office.

In the mid 1990s, I started a real estate publishing company, your typical free real estate advertising magazine. This was very early internet time and I created a web site and put the homes on the internet. Fun stuff, but not very profitable. Developing the web site helped me land a job with Fidelity Associates.

I worked at Fidelity Associates as a software developer from 1999 until 2015 and retired 2/27/2015.

I married my wife Celine December 15, 2007. I'm a very lucky man. We live in my childhood home, upstairs, and my 99-year-old Mother lives downstairs. We are her caretakers, although she is very independent and does most daily activities on her own. We are planning a 100th birthday party this coming April 8th. Mom is very excited and looking forward to the party—she's a real treasure.

Celine and I recently bought a winter home at the Plantation in Leesburg, Florida where we plan on spending 6-8 months of the year. This is our first year in Florida and being close to Universal Studio and Disney World and are looking forward to my daughter Tess, along with her husband Wes and our four grandchildren—Connor (14), Caleb (13), Cohen (11), and Cooper (9) coming to visit us.

My wife Celine has two children and a grandson who is attending American University in Washington, DC. Her daughter Kathleen lives in Vermont and her son James lives in New York. My lovely wife will soon become a Nana as her son James and Sarah are expecting a baby in May.

Enjoying life and retirement.

BLECKERT JR., GORDON E.

COWBOY-23 (MUSTANG) 1,351 combat flying hours.

After Vietnam, the Army sent me to the 349th AV. CO. Ansbach, Germany. Requirements were top secret (they even found my first-grade teacher).

Discharged 1971 Brooklyn Army Station Civilian post military. I then took several months living leisurely.

Became employed by an office furniture manufacturer. Spent time working in all phases of production and became the quality control director.

I was then recruited to become quality control manager of large home goods (i.e. Furniture and Workrooms) at May Department Stores, Cleveland. Promoted to establishing, along with another person (civilian trained pilot), an industrial engineering department to concentrate on material handling from receipts, vendors processing, and movement stores selling floors. I was then made Manager of Material Handling after establishing work flows and standards. Remained in that capacity several years.

To advance my career in retail, I moved to store management as operations manager of a medium size store. Then moved, after a period, to the divisions largest volume and flagship store.

I was offered a similar position at The Broadway Stores in southern California and I accepted the position. During a tour of the other stores in the region, I was talking to the director of personnel at the Glendale store, Alexandra Kelly, and several months later we had our first date. This led to a fantastic relationship and marriage. Alex and I are still happily married.

Left the Broadway Stores for the construction business. Briefly, this was a great experience, the people I was able to meet and organizations I got involved with were exceptional. Alex remained with the Broadway Stores and moved into the corporate personnel ranks. Macy's Department Stores purchased Broadway Stores and evaluated the staff and chose to retain Alex and offered her a position as Vice President at the corporate headquarters in Cincinnati, Ohio (yes, Cincinnati). This would allow her to become a VP at a major corporation before her 40th birthday. I agreed and ceased my activities in LA.

Alex is currently a group vice president with Macy's, she's not 60 yet (11 years younger than me) and really loves what she's doing. I'm retired and as I've always enjoyed golf, am still doing so.

When Alex does decide to retire, we will be relocating to the Carolinas.

BOLIVAR, ANTONIO A. (TONY)

- First Tour Vietnam, Casper Aviation Platoon, April 67-March 68.
- Fort Wolters TX March 68-October 70 Instructor.
- October 70-December 70, AMOC, CH-47 transition.
- January 71-December 71 Vietnam 610th Transportation Co. Da Nang.
- January 72-June 72 Fort Wolters. ETS.
- June 72-January 74 University Texas Arlington.
- February 74-January 93 Aerospatiale Helicopter Co.
- June 93-June 2012 Parkland Memorial Hospital Dallas Texas. Retired Director EVS.
- June 72-October 87 Texas National Guard, 336 Helicopter Co. CH47 (Hook Masters) Retired CW4 November 87.
- Married 48 years, two children, four grandchildren.
- Live in Arlington TX. Belong to AMVETS, Casper Aviation Platoon.

BONNELL, TERRY W.

After I came home from Vietnam I had fourteen months left in the Army. I was stationed at Fort Lewis, Tacoma, Washington for the whole fourteen months.

When I was discharged from the service, I came back to New Philadelphia, Ohio. I married my wife Linda between getting home from Vietnam and before I went to Fort Lewis. We have two children, a girl Tara and a boy Todd. We also have four grandchildren.

I started looking for a job. I found work on the 3rd shift for about three years in the plant. I applied for advancement to Quality Control Inspector. I decided to get my associate degree from Kent State. I managed to get two associate degrees. I worked there for about ten more years before being laid off. I was out of work for about two years.

I was once again looking for a job, so I answered an ad for the government as a Quality Assurance Specialist and got the job. I then moved to Toledo, Ohio. I was there for two years when I transferred to Cincinnati, Ohio. I was there until January 31, 2014 when I retired at the age of 66.

I belong to the Schoenbrunn Moravian Church in New Philadelphia, OH. I'm a life member Vietnam Veteran Association Chapter 857 there in New Philadelphia, OH.

BOTHUR, THOMAS JOHN

CW5 (Retired) Mustang 23

My true love has always been to fly. I have been flying for over 50 years. I have always enjoyed viewing the world from a high perspective. As a kid, you were sure to find me climbing trees, clambering up hill tops, or jumping off our roof, using a sheet as a parachute (which much to my dismay, did not work.). Flying was so interesting to me that I joined the Civil Air Patrol.

In 1968, while I was still attending high school, I heard about the deal of a lifetime. It was promoted by Uncle Sam, that if I get my diploma and pass a few tests, he would afford me my dream of being able to fly. With urgency, I passed all the tests. It was confirmed, I was not only healthy but nuts enough to fly without being too crazy for the Army.

On a fine August day, it was off to basic training at a lovely place called Fort Polk, Louisiana. After being inspected, injected, indoctrinated, and subjected to all the other routines and rituals, I emerged as a lean, mean fighting machine! From there I was driven to Fort Walter, Texas for the initial flight training.

After completing the initial flight training, I was sent to Fort Rucker, Alabama for advanced training. Once I completed this training, I was reborn as an Army Warrant Officer and a helicopter pilot. It was now 1969 and Uncle Sam had waited long enough. It was time to collect for all that training and nifty new flight suits they had provided me. It was off to a foreign country half way around the world, to a place called Vietnam.

The first leg of the journey was on a commercial flight from the east cost to the west coast. Being under the official drinking age, I had to get the pilot involved to get a proper drink. After all, I was now officially a helicopter pilot going off to war. Bless the pilot who ordered the stewardess to serve me.

During the war, my job was combat assaults, medevac, resupply, and troop extraction. I also flew Nitehawk and counter mortar missions. My pucker factor was high without the aid of night vision equipment. Regardless, we were successful in our part of giving the bad guys a bad end to their day.

After a year of war and what seems like a lifetime of stories to tell, it was back to the land of the big PX at Fort Dix, NJ. Upon arrival, we were told to get out of our military uniforms before leaving the base for home. After all it was the 1970s and the anti-war movement was still in full swing. I was assigned to Fort Meade, Maryland. There I flew OH13 and UH1 aircraft. Back then you could fly over the monuments of downtown DC and places like Gettysburg without being intercepted.

By March of 1971, our involvement in the war was being scaled back and the Army was downsizing. I was offered an early out from active duty and took it. I was not ready to leave the military, however. I joined the Connecticut National Guard the same day that I left active duty. I was released from active duty in the morning and sworn into the Connecticut National Guard that afternoon. I was the first Vietnam Veteran in Connecticut to join the Guard.

Throughout the 1970s, 1980s, and 1990s, being part of the Guard family was a safe place for Vietnam Veteran's. It became a home away from home with a team of people you could trust and confide in. It was a place where those around you shared the experience of combat. It helped keep the bad out. It was therapy without seeing a shrink. What I miss the most is the comradery with other men who served that would have risked their life for you, no questions asked, as I would for them. Who else besides a Veteran could say it does not matter why we were at war, what was important is who was covering you on your right and left.

Luckily, I had some equipment from my time in active duty, as the Guard was still operating on a surplus of World War II and Korean leftovers. Our helicopters were small observation ones like the ones seen on the television show MASH. We had a few larger Korean War Sikorsky and two little fixed-wing aircraft, too. It was basically a flying club with no real-world mission until 1973 when we received our first Huey, an old B model. As more Vietnam vets joined and the other guard members became qualified to operate a Huey, we started to receive more surplus helicopters. It was a long and tedious process because we were at the end of the supply chain.

At times, the Guard suffered from serious funding issues and we only had enough bullets to fire qualification rounds without enough for practice rounds. Our 45 pistols were so worn they rattled when shaken. It was a choice at times to pay the Guard members or gas to fly the helicopters. Grounded due to budgetary concerns; a Guard pilot's worst nightmare.

There were years where we had our scheduled annual training canceled and held locally because it was too costly. We were supplied with World War II and Korean War leftover rations. I recall one of the cooks opening a can of hot dogs that had turned green. Thankfully, we were blessed with thoughtful and talented cooks who took great care of us. They would sort through the crap given to them and pick out the usable goods. Sometimes they would use their own money to go to the local market and pick up other things to create a more than edible meal. Talk about cooking with love.

Despite the lack of proper financial backing, we built a flying program that became a model for other States. Through constant lobbying at the Guard Bureau, equipment found its way to us. In the early 1980s, we had over 28 aircraft. Our troop strength was over 300. At this peak, we were faced with funding and man power cuts as peace time swept the nation. Unfortunately, we lost two-thirds of our manpower and were down to 100. What used to be a platoon, was now a company.

When 9/11 happened, everything changed. There was new equipment, ample training funds, and a public that supported its military. I was very uncomfortable when people started to thank me for my service after all the years of being ignored. When I was deployed in 2005 to Iraq, I was amazed at all the equipment and other assets that came with it. While in Iraq, if we needed a major component, like a new aircraft transmission, it would be to us within days. Back at home that would have taken months. Currently, it appears that the cycle is repeating itself. Everyone is looking for new peace dividends, causing a roll back of troop strength, cuts in training funds, and equipment.

On 9/11, I left my work even before the second plane hit the towers and headed to my unit. We were one of the first guard aviation units to respond. I flew into New York City to bring in personnel, supplies, and conduct aerial surveillance. It was after the third day when it became a recovery mission that I was released. Flying back to my base through what was always a busy airspace, not a sound was heard on the radio. When I questioned the flight controller, his response was, "You're my only traffic."

My career outside of the military started with being a machinist and avid union member. I also worked my way up to Executive Director for a non-profit company. I utilized my G.I. Bill and received my BA degree in business which helped achieve the top position. I was the first of nine children in my family to obtain a college degree. While still on active duty in Iraq, I decided to fly commercially instead of going back to my desk job. I currently still fly an S76 for the rich and famous.

There are many things that I have seen or done during my flying career. While in Vietnam, there were Huey's with ten troops and four crew bouncing off the ground to get airborne. During Iraq, there was a Blackhawk with twelve fully loaded troops and four crew in 100 plus degree temperature getting off with power to spare. Flying at low levels during nighttime with only your eyes and then learning a whole new way of flying, using night vision goggles down in the trees.

Flying helicopters where I had to coordinate every movement, cyclic pedals collective (human computer). Helicopters that have mechanical mixing units to do that and helicopters that have autopilots that almost fly themselves. At least there is a button to take the computer out, so I can regain some dignity as a pilot instead of being a systems manager, I use it often.

Both in the Guard and as a civilian pilot, I have searched for lost people and downed aircraft. In my current job, I have flown in Santa for scheduled events.

I used to do candy and toy bombs for the kids in Iraq. The trick to ensure the little ones got their share was to drop a soccer ball so the bigger kids would chase them while you fly over to the smaller kids and drop stuffed toys and candy to them.

I have been married twice. I married my first wife before Vietnam. We have two wonderful children, one boy and one girl. We divorced not long after Vietnam. My marriage ended because I had developed a lack of empathy for things. It was hard for me to view certain life events as seriously as others saw them. After experiencing war, life and death issues had a whole new meaning. I eventually remarried, and we share yet another wonderful son. That marriage ended after 16 years.

I enjoy sharing my love of flying with others at air shows, in a classroom setting, or while assembled around my aircraft. I have given speeches about my time in service and asked to be the keynote speaker at Veteran's Day events. When I spoke at my town's Vietnam Memorial dedication, there were two other Cowboys' in attendance, which made it very special. It is such a spectacular honor to do flyovers at events, especially at Veteran's funerals. Presenting the flag that covered the casket of a friend to a widow has always been a difficult, yet privileged, honor.

There are things I currently enjoy doing. Skiing is a fun thing for those cold winter months. I let gravity pull me down the hill, of course, too much work cross country skiing! Golf is a good time to enjoy the outdoors, having a drink and a cigar. I don't take it very seriously yet, I'll do that when I'm retired. Traveling to other places where there is no active war, though war does make for some interesting times. Also, it is fun flying over a herd of camels, seeing old temples, and the ruins of the beginnings of civilization. I like to ride my Harley motorcycle, my kids joke, "There goes dad on his bike to go fly a helicopter."

It is also fun teaching new groups of people how to stay alive in a war zone, like I was taught by the COWBOYS. Land with your tail to the bad guys, not the nose, don't fly over the same route, don't cross power lines at the same spot going into a base. Fly fast and low when you can to minimize exposer time. Brown outs in the sand box are just like whiteouts in the snow. Remember the objective of this game is to bring everyone home with you. Just same shit, different day.

I will not be remembered for being a good soldier. My first OER in the National Guard said, "He is a good aviator...He is not an outstanding soldier...Probably will not make a career of the military." Not only did I stay in the military, I retired after completing 41 years as a Chief Warrant Officer 5. I still would not make a good soldier! I am by all accounts, one hell of a pilot.

I brought home a souvenir from my time in Vietnam. It is likely derived from a chemical called Agent Orange. In April of 2017, I had my right middle lung surgically removed. The doctors said they got all the cancer. The FAA and flight doctors willing, I will continue to perform death defying acts of flight and see this great world from a high perspective!

Thanks to all the support (crewchiefs, gunners, maintenance, supply, and fuelers) people who have made my flying career a long and safe one.

I am proud to have been a member of the COWBOYS.

BOWEN, ROBERT (BOB) E.

I was drafted in March 1965 from my hometown of Marion, Iowa. Went to basic at Fort Leonard Wood, MO. AIT was at Fort Rucker, AL. MOS was single engine, single rotor helicopter mechanic. Met my future brother in-law Tom Buser in basic, stayed together through AIT and over to Vietnam in September 1965 to the same outfit, the 166th Transportation Detachment in Vung Tau. A couple of months later, we went to Bien Hoa. I joined the Horsethief crew and Tom ran the Playboy club, our enlisted men's club in the 166th area. Our first CO was Captain McConnell, later Captain Prichard. I flew with Sam Bailey, Joe Fields, Norm Usher, and Matsumoto.

Best experience was going on the Playboy Bunny flight, taking Jo Collins to visit the troops in various locations.

Worst day was my birthday, July 27, 1966. We were accompanying the Cowboys when one of our slicks had a mechanical failure and crashed. All aboard perished, four crew members:

W1 Rutherford Welsh
W1 Joseph Sampson, Jr
SP5 Harold Reinbott, Jr
PFC James Collins
and six members of the 503rd Infantry, 173rd ABN.

My last few months in the service were spent at Fort Benning, GA. After being discharged in March 1967, I went to work at Collins Radio Company in Cedar Rapids, later it would become Rockwell Collins, a maker of avionics equipment. I worked there 35 years, eventually managing a Computer Aided Design Center. I met my wife

Diane at Collins Radio, she likes to tell people she trained me in my first job there. We will celebrate our 50th anniversary on October 19, 2018. We have two children, a daughter Christine Slump (husband David, is a former teacher) and a son Rob Bowen, (wife Kim) works at Rockwell Collins as a Computer Engineer. They each have a daughter and a son. The two granddaughters are sophomores in college and the grandsons are sophomore and junior in high school.

After retiring, I have spent my time working on and driving my 1967 and 2003 Corvettes and building a 1949 five window Chevy pickup for my grandson, who is helping me on the project.

My military experience has made me a strong supporter of this country and of those men and women who serve to protect us. Every time I visit the VA hospital and see all the young men and women who sacrificed so much more than me, I feel so very blessed. All I lost was some of my hearing and got Type 2 diabetes. Small price to pay compared to those that gave it all, or part of themselves.

I think the military very much affected my life and how I view things and what I respect.

I belong to American Legion Post 298 in Marion, Iowa and to the Cedar Rapids Corvette Club. One of my proudest accomplishments as a 50-year member of that club was to go on the Honor Flight as a guardian for a World War II veteran. Our club has provided a commemorative plaque to each veteran on the Honor Flight, to date over 2,300 since 2009, when the flights started out of the Eastern Iowa Airport in Cedar Rapids, Iowa. In April of 2018, I will join about a dozen of my club members who are Vietnam veterans as we take the Honor Flight to DC.

In 2003, with our wives, my brother in-law Tom Buser and I attended the Cowboys reunion in Las Vegas. Had a wonderful time, met a lot of new friends, most of the guys there had joined the Cowboys after I left in 1966.

In 2004, my wife Diane and I took a cruise to Alaska where we met up with fellow Horsethief crew member Joe Fields and his wife Cindy at Denali Park. After leaving the Army in 1967, Joe went back home to his wife in Fairbanks, Alaska. We had a wonderful time that day. We had not seen each other since I left in 1966, but we had been communicating by phone and later email.

While viewing a post on the old Cowboys website I saw where Cowboy George Murray lived near my daughter in Roswell, Georgia. Since then I have had coffee with George a few times while visiting my daughter. It was during one of these visits that I found out George and I share the same birthday, July 27. Although George had that birthday a year before me. George was on that same flight when we lost the Cowboys on our birthday. He lost one of his good friends that day.

The summer of 2010, my wife Diane and I traveled to Mineral Wells, Texas to visit my former CO Captain Mac (Demer McConnell). I had been communicating with Captain Mac via email for many years after locating him. We had a great dinner and we visited until after 1 AM the next day. Captain Mac was so glad to see one of his "boys" and I was so very happy that we had made the trip that I had been telling myself for years to take. Less than a year later (July 23, 2011), Captain Mac passed away at his home in Mineral Wells, Texas. I'll always cherish that day we had together in Texas.

If you ever get a chance to meet up with some of your fellow Cowboys do it, don't put it off until it is too late.

I have been blessed with a great wife, great children, great grandchildren, super family, and super friends and most of all my Cowboy friends that I am lucky to have.

God Bless all of you.

BROWN, EUGENE PATRICK

Prior to entering the military—I graduated from High School in 1966 and attended a computer technical school in Pittsburgh, PA for nine months. Obviously, when I completed my schooling with a 1-A draft status, I was not able to obtain a job. So, join one of the branches of service, or be drafted.

How did I make it to the 234th Signal Detachment, assigned to the 335th AHC? In 1967, I was about to join the Navy, but at the last minute I changed my mind and volunteered for the draft. The draft board informed me that they had already drawn the names for the coming month, but I just asked them to put someone's name back and draw mine. Two weeks later, I received my draft notice. Basic training at Fort Jackson, MOS training (Radio Repair) at Fort Gordon, and then off to the 335th at Phu Hiep in February 1968. We relocated to Bear Cat somewhere around the end of year 1968. I spent a month or so at Bear Cat. I repaired the UHF radios and associated gear (helmet cords, mics, etc.) each night as the Cowboys returned. Once I finished my tour of duty, I had less than five months of time left of my two years, so they did not reassign me. I was discharged with serving only twenty months. And since you still had a six-year commitment to the service, I had to wait for my final discharge. And wouldn't you know it, even as a Vietnam Veteran, I was called to go to summer camp during the two years of active reserve. They assigned me to attend summer camp with an active reserve unit in a neighboring city. And when I showed up in my Vietnam fatigues, they all asked what I was doing there. Of course, they tried to recruit me into their reserve unit, with no luck. Quite an experience playing war games with a reserve unit.

My military experience was not in line with my computer technical training that I received prior to entering the service so I reverted to my computer training for my civilian life. The only use my military MOS (Radio Repair) training served was to inspire my hobby of electronics.

After Vietnam, I married my high school sweetheart, Vickie, and started my first real civilian job at Mellon Bank in Pittsburgh, PA.

We were married in 1969, one month after arriving home from Vietnam. We were blessed with two children and five grandchildren. We spent all our life living in Pennsylvania. My career started with 23 years at Mellon Bank, Pittsburgh, PA and 20 years at various companies in the Harrisburg, PA area. In 2012, I retired from the Information Technology industry after 43 years.

I attended the University of Pittsburgh using the G.I. Bill, working full time while attending school full time in the evenings and eventually earning my BS degree in business.

After I retired in 2012, we relocated from the Harrisburg area back to Pittsburgh where one daughter and three grandchildren live (and where we raised our family). Our second daughter lives in the Seattle area with our other two grandchildren.

I am currently a member of the VFW and American Legion. And as a retired Mellon Bank employee, my wife and I spend a lot of time volunteering at many charities, blood drives, etc. as well as volunteering for the 412 Food Rescue, where we rescue food from various grocery stores and take them to charity organizations.

My military experience has not been very influential in my life. I've been healthy with no signs of any PTSD or Agent Orange impacting my life style. I do believe that going into the service made me grow up in a hurry, rather than going the college campus route. Since I recently watched the Ken Burns Vietnam documentary, I have a deeper hate for politicians and how they are dishonest with the American public.

CHANDLER, RICKEY E.

Civilian Education

- 10 + semester hrs. Community College
- 66 hr. Tee Institute, Beginning Arc Welding
- 5-year apprentice plumbing and heating, Refrigerant Transition and Recovery Certification (universal)
- NC State Contractor Plumbing License

Military Education

- RC-PLDC 2 Weeks
- RC-BNCOC PH 1- 2 Weeks
- BNCOC PH 2-2 Weeks,
- Leadership Course (10 hrs. correspondence)
- Infantry Advanced Refresher Training Course (MOS 11E)
- 108 hrs. correspondence Enhanced Remoted Target System (ERETS 80 hrs.)
- Decorations, award and citations (spelled out in order of precedence):
- Combat Infantry Badge, Basic Aircraft Crewman Badge, Air Medal with 25 OLC, Army Commendation with 3 V-2 OLC, Army Achievement Medal, Army Good Conduct Med 4, Army Reserve Component Achievement Medal with OLC, National Defense Service Medal, Vietnam Service Medal,

Armed Forces Reserve Medal with SING, NCO Professional Development Ribbon with 2, Army service Ribbon, Republic of Vietnam Campaign Medal, NCNG Meritorious Service, NCNG Commendation, NCNG Service with GM, NCNG Governor's Unit Citation, NCNG Meritorious Unit Citation, Basic Parachutist Badge, US Army Distinguished Rifle EIC, US Army Distinguished Pistol EIC, Chiefs 50 Winston P Wilson, OCTOBER, 1997, President's Hundred AUGUST 1999—AUGUST 2000.

Civilian Affiliations

- NCNG Assoc. (life member) AARP, NRA (life member), Association of Marksmanship in the National Guard, North State Shooting Club, North Carolina Rifle & Pistol Association, Buggs Island Striper Club, Inc.

COCKRELL, DANIEL DAVID (DAVE)

After Vietnam I was sent to Germany where I finished my tour of duty with the Army, then I came back home to Bloomfield, MO.

One of my best memories of Vietnam was when Major Gammons came to the flight line and gave me my stripes for an E-5.

After leaving the service, I returned home and worked as a tree trimmer for an electric company. I also attended welding school, as well as breaking and doing some training of horses on my own.

In 1972, I married and moved to Caruthersville, MO where I worked as a welder at the Shipyard. While we lived in that area we had three children. One boy, John David, and two girls, Marsha and Teresa. I always enjoyed the cowboy way of life that I had grown up with, so I decided to move back to my home area and work on a ranch. I worked with cattle and hogs as well as breaking and shoeing horses. While I loved the work, the pay wasn't enough to raise three children on, so I went to work at Lowe's Clay Company as a mechanic.

I continued to break, train, and shoe horses as a secondary income, and with my boss at work also being a horseman and raising quarter horses, he would allow me to ride the horses that I was working with to and from work. I lived three miles from work, so it was great in riding out the horses that I had broken.

I worked 33 years for the same company, but during my tenure there, the company changed names four times. It started as Lowe's Clay Co., to Southern Clay Co., to Golden Cat, to Purina, to Nestle-Purina Pet Care. We manufactured the "Tidy Cat" Kitty Litter.

In staying with my love of "the cowboy life" I had two significant trail rides. In 1976, my brother Richard, my cousin Jerry Wheatley, and myself made a bicentennial trail ride from Hayti, MO to Bloomfield, MO that took about three days. Then in 1983, there was a group of four that made a trail ride from Bloomfield, MO to the Lake of the Ozarks and back. It was a 600-mile trip that took sixteen days. I shoed the horses we rode, and I was the only one to complete the whole trip. It was a wonderful experience.

In 1995, I divorced and have never remarried. I bought a farm while I was still working and raised and sold cattle for several years.

In 1999, at the age of 49, I had a heart attack that resulted in having open heart surgery. It was later determined that Agent Orange was a contributing factor to my heart trouble.

In 2009, my only son was killed in an automobile accident in Wyoming where he was living at the time. I arranged for him to be brought back home to be buried next to my parents. Not long after his demise, I sold the cattle and farm. John and I spent many hours on the phone together when I was at the farm and due to some problems with PTSD, I was not able to cope with going back to the farm and taking care of it. So, it was best for me to sell.

After retiring, I still live in Puxico, MO. I have some land and I continue to fatten out calves and hogs, and I continue to raise my own garden. I have chickens, ducks, cats, dogs, and of course I still have horses. My health is great, and I am enjoying life.

I plan in July to make a trip to Montana to see a great friend from Vietnam, Rick Geary.

To sum it all up, I was a "Cowboy" before the Army, a "Cowboy" during the Army, and a "Cowboy" after the Army, but I can say that I have a Government Certification to document that I, Dave Cockrell, am a Cowboy! LOL

CONDRY, DARRELL WILLIAM

1SG Retired

After serving with the Cowboys from October 1967 to December 1969 in Vietnam, I returned to the USA (as a SP/5) and was assigned to Fort Rucker, AL. They had so many UH-1 crewchiefs there they didn't know what to do with them. Ron Trouard was working in the Engine Run Cell, Russ Stibbe was being assigned as an Operations Specialist at Lowe AAF. I ended up as the Ops Sgt for the Attack/Scout IP Branch. The pilots taught others to become instructors for the M22 Missile. While there I went home and married the young lady who was writing to me in Vietnam, Nancy McClelland. A year and a month after arriving at Mother Rucker I found myself taking Nancy back to Maryland and on my way to South Korea. Our first child was on the way, too, so it wasn't a happy situation.

I arrive in Korea February and ended up as a maintenance crew leader in the 337th Air Medical Co. Our daughter, Nanette Eileen was born in April. While with the 337th I hadn't counted on hearing AK 47s again, but I did. I had to go into the edge of the DMZ to recover one of our birds that the north had shot down. It had gone in to pick up some ROK guys who ran into a NK patrol. The bird took a hit on a Xmsn oil line. It just made it back inside the border, but the fire fight kept on.

During our annual IG Inspection, the Inspector came into my room and asked me if we had a drug problem in the unit. I told him we sure did. He walked to the end of the bay area and turned down the stock gated mattress on an empty bunk. There was a small bag of weed in the mattress. He turned to the Commander and said that at least you have one honest NCO. Time went quick, there was no slow-down in the action.

I PCS'ed back to the USA, gathered my family, and we headed to Kansas. Surprise, Surprise, Surprise. I was assigned to the 335th AHC, just returned from Vietnam and was reforming. The Cowboy sign that stood in front of the orderly room in Vietnam was in front of it again at Fort Riley, Kansas. I was assigned to the Falcons, but this time we had Cobras, AH-1Gs, six of them. Four used and two right off the as-sembly line. I received one of the new ones. It was a lot different than Vietnam. I was promoted to SSG. Three years later, I packed my bags and headed back to Korea. Nancy went back to Maryland to wait.

I ended up in the 2nd Infantry Division, with E Co. 2nd Avn Bn. A support aviation unit. My time with them was short as the Platoon Sergeant position came open in B Co. 2nd Avn for the Cobra Platoon. I interviewed and received the job. I was also assigned as the lead Tak Won Do instructor for the Battalion. I had a belt in it from my ear-lier assignment and everyone in the 2nd ID had to have six weeks of defensive TWD instruction. A lot of things were happening in Ko-rea at that time. I ended up spending an extra year there. That's a sto-ry in itself.

Next stop was Fort Hood, Texas. I was assigned to B Trp. 7/17 Cav., 6th Air Cav Brigade. I went from being a Maintenance Super-visor to Tech Inspector to a Platoon Sgt then back to Inspector. This was when the Army went to aircraft specific inspectors. I was the only inspector trained in all three of the units' birds, AH-1S, OH-58A and UH-1H. Where the aircraft went, I went. Sometimes I flew with the Commander and sometimes front seat in a Snake. We trained from Texas to California. A little over three years later, I was headed to Ger-many. Nancy and the girls (Megan Diane, daughter #2, was born at Fort Hood) were to follow when I secured quarters.

Upon arrival to the 8th Infantry Division, (Mainz-Finthen) I was assigned to B Co. 8th Avn Bn as a Quality Control Supervisor. We had three Flight Platoons, each one consisting of seven Snakes and four Scouts. We also had three UH-1Hs for support. We supported the 8th ID and we had a defensive sector (GDP) on the East/West Border. I received temp-quarters and Nancy and the girls came over. The airfield was about a five-minute drive from our quarters. A few months later we moved into permanent quarters a few buildings over. The quarters were very nice, and we enjoyed them. We had alerts about every six weeks. All were timed, and we had to be packed and ready to move out to our GDP area. I ended up taking over a flight platoon. The Battalion Commander and I had a difference of opinion on some issues. It got downright nasty at times. I survived! I made E-7 and asked to have him pin my stripes on, he did. When he rotated out, he said the next time we met he expected me to be an E-8. We flew all over Germany and other parts of NATO. I lost two great pilots when my maintenance pilot and SIP where killed on a night training flight. I had all my crewchiefs go to the crash site to see the aircraft, I wanted them to see what could happen and understand the importance and responsibility their job carried. It was determined that nothing they had done caused the accident. We left Germany not only with a clock but with our third daughter, Katherine. We adopted her, and she has her own story!

Next stop was Fort Rucker to the Department of Enlisted Training. The three branches it was responsible to train were Flight Operations, UH-1 Maintenance, and OH-58 Maintenance. I was assigned to the OH-58 branch. My boss was the Branch Chief of the OH-58 school. I was assigned as the Chief Maintenance Instructor. We had four levels of AIT training. I sat in on classes to insure the training was correct. The students had to test out of each level. Any test failures, the boss or myself had to counsel and either put them thru the level again or recommend reclassification. We had DA civilians and foreign students go thru the training course.

The same day the Challenger blew up, Nancy's dad passed away. We went home for two weeks and then back to Rucker. I also took my over 40 physical and was flagged for failing to recover properly after running. EKG indicated possible heart blockages. Off I went to Walter Reed, yep I had three small ones and they put stents in them and returned me to duty. While at Rucker I used some of my Education Benefits to attend Alabama Aviation Institutes night school. The course was General Aviation Technology. It prepared me for the FAA's Airframe and Powerplant Technician License. I passed the tests and received the license. I also was promoted to E-8. I applied for retirement and it was turned down. I was on orders to the Apache Training Brigade. I had the time left on my enlistment to do it so off we went.

When we got to Fort Hood to the Apache Training Brigade, it was unlike anything I had been through. They had their own housing, my family looked at several houses, we picked the one we wanted. They had two Apache units forming up, I had my choice of which one I wanted, it was unreal. I ended up as First Sargent of B Co. 2/227th Advanced Attack Helicopter Bn (Death Dealers). It was a challenge. My Commander was decent, but it was the melding of A.I.T students, both enlisted and pilots into one fighting unit. Officers had their ideas and some enlisted had theirs. With tact (I threw my Kevlar Helmet at the Attack Platoon Leader while in the field once), and demonstrated knowledge of what to do, I usually won out. B Co. excelled in all the tasks we had to perform and in August we left for Hanau, FRG, families and all.

Three days after arriving in Hanau, Germany, someone knocked on my door at 2:00 AM. It was my Commander and the S-2 Officer. B Co. was being further deployed to Saud Arabia for what would be Desert Storm. The next morning, I was told to report to the Battalion Commander.

I was not deploying with my company. The men and aircraft were going to plus up a sister Battalion that was going. I would stay behind to assist the families moving into permanent quarters, picking up vehicles, and receiving household goods. It was an extremely heated discussion between the Colonel and me. I lost. So, off my guys went. They were there several months and then returned. This time everyone was going except, of course me and two others, an E-4 with medical problems and an E-7 who would be doing courier duty for the Division (3rd Armor). While the Battalion was getting ready to go, my mother passed away, my family went stateside with me for two weeks, and then back to FRG. A lot transpired during the Battalions deployment to Desert Storm. That is another story all by itself. When the Battalion returned after Desert Storm, I immediately asked to be transferred. Too much had happened during its deployment and I had to go. I was transferred to Headquarters and Headquarters Co. 227th Avn Bde. There I was assigned to the Installation Command. And that adventure is a whole chapter in itself.

I was caught up in the Army retention change by rank about a year after Desert Storm. I retired 1.5 months short of 26 years. We returned home to Maryland. I worked for a State Summer Work Program for city youth as a supervisor. Our work area was Green Ridge State Forest. We cleaned up forest areas that were harvested, cut hiking and biking trails, improved camp sites along the Potomac River, and anything else the Forest Rangers wanted done. After that, I worked for a heavy aircraft maintenance company working on Boeing 727s. I applied as a Helicopter Tech with the Maryland State Police Aviation Command and was hired. I worked as a Tech on Eurocopter AS 365s. Later I received a Master Tech rating from America Eurocopter Company. My next goal was to take and pass the FAA Inspection Authorizations Test. I did and became an Inspector for MSP. Next, I did a stint as their Director of Maintenance.

After almost 18 years driving a minimum of 200 miles a day, I retired from MSP and went to work for Mecaer Aviation Group (MAG) as the Chief Inspector, 15 miles from home. I wrote the Operations and Training Manuals for the Repair Station and worked with the FAA to get it a FAA Certified Repair Station. Mr. Ed Pears was the company manager. A great guy to work for. According to the FAA's Baltimore Office, it was the quickest standup they had seen by any company to date. We performed Helicopter Maintenance and VIP conversions. The group had two work sites, Hagerstown, MD, and Philadelphia, PA. The main site was the one in PA. MAG was an Italian owned company. After three years of working for MAG, Mr. Pears suddenly passed away. MAG made the decision to downsize to just the PA site. I was not willing to move to Philadelphia, although I did some fill in work for them later that year. After that I was ready to retire and so that is what I did.

Looking at some tips we were given to get this story:

Did you work or go back to school? What did you do as a career after Vietnam. I stayed in the Army, and by doing that I worked and went back to school in my career field. I travelled across this wonderful country of ours. I not only increased my military training but training in the civilian world of my chosen career in Aviation Maintenance. My older brother who was a retired career soldier (Army), and served two tours in Vietnam, used his education benefits for a career in computer programing. He passed away several years ago, Agent Orange got him. My younger brother also served in Vietnam with the US Marines, used his for a career in banking.

Looking back on the military, there are good things about it and some that are just bad. The military after Vietnam wasn't a good place to be. There was a lack of trust between the Officer Corp and the NCO's. I found myself having to prove to them that I knew my job. It would take several months and sometimes longer to get them to trust me to know what I was doing. They had this thing about only trusting themselves. Where was this being taught them? I know that this was the case with almost every 2nd Lt. coming out of West Point. In Vietnam, the flight crew was just that, a crew. We trusted one another to do their job. If the pilot wanted to dive on a VC gun emplacement, he did or fly thru whatever to resupply or pick up wounded grunts, he did. We were a crew. We trusted in each other. I only hope that the trust has returned in one another. My military experiences taught me that life is precious. You do only go around once. You must stand on what you believe. That is how I have lived my life. I am comfortable in my own skin.

Several years ago, I had to have a letter from one of the "Cowboy" Commanders to verify my character while I was in the unit. I rotated back to the States with a basic "Air Medal." That might not seem like a problem unless you are a career soldier. I sent a certified copy of my flight records to DA, trying to get my other medals. I had to prove I was of good character. Then they still only gave me a couple more. They said they didn't know how they were awarded during that time. I also submitted a copy of a medical report where I took a nose full of shrapnel when some rounds came up thru the floor of my aircraft. After I retired it took three surgeries to repair the damage in my nose so that I could finally breath right. Again, I couldn't prove I was of good character. You have to be able to laugh! My life has been a soldier's life. I'm proud of it and the duties I performed. I will always have pride in the Cowboy organization and the people I have been lucky enough to serve with!

I will also say that I have a 90% disability rating from the VA. Agent Orange is something that should have never been used in Vietnam. It has poisoned all who served there and continues to do that today to the people of Vietnam.

I am a life member of the VFW.

CORB, RONALD M.

When I left Vietnam in December 1970, I went home for a 30-day leave. Had a good time. I had a lot to catch up on and a lot of steam to blow off. I wanted to get it all out because I was being sent to Germany for a year.

My whole time in Vietnam I was a door gunner, never a crewchief. In Vietnam, I had a good friend that worked in the orderly-room who changed my MOS from door gunner to crewchief. While home on leave, I received orders making me an E-5. The funny thing about that is when I got to Germany and reported to the 350th Aviation Company, the first sergeant made me a maintenance platoon leader, because I was a Vietnam Veteran and a crewchief. Little did he know I knew nothing about the workings of a UH-1 Huey. And I didn't tell him. It worked out for me because I had a good crew. I had a lot of fun in Germany and I even got to hook-up with Smitty (Robert Smith) for a few days. We had a lot of fun drinking and a few other bad things.

After a year in Germany, I came home and took off for about a month before I started looking for a job. Took almost four months before I finally found one at a washing machine plant. Then I did what every young guy does. I bought a new car. A 1972 Dodge Charger 400 Magnum, beautiful and fast.

Sometime after that, I received a call from Doug Wilson. Took off for two weeks and went to California. We spent most of our time in Mexico having a good time.

My job only lasted a little over three years. I was fired because I refused to work overtime. I think the real reason was because I was trying to get the union in the plant. Didn't worry me. I was running a card game and working in a pool hall. I made more money doing that than working a regular job.

In 1977, I knew I had to get a regular job because I had plans to get married. I finally landed a job with a beer distributor. Worked there for the next 35 years. During those years I played a lot of softball and football. Also went to a lot of NASCAR races and Indy 500.

In 1991, I received a call from Doug Wilson about meeting up with some of the guys from the 335th. I was all for it. In 1993, we had our first reunion and it was a week to remember.

When I retired in 2012, I was a happy camper. After moving those 160 lb. kegs around, anything else I did was a piece of cake. Of course, it kept me in decent shape. In January 2016, I came down with Type 2 diabetes. Blamed it on Agent Orange but it probably was because I got fat and lazy. I quit drinking soft drinks and eating a lot of candy. Working out four days a week also helps.

Don't really have any hobbies, unless playing cards count. Still married, have three good kids and five grandkids. Looking forward to our reunions keeps me going.

CURRY, MICHAEL T.

I joined the Army in 1966 at the age of 17 in Portland, Oregon. I had quit high school due to a car accident injury and would have had to repeat a grade. I didn't like that idea and knew I would probably soon be called up so thought I may as well get it over with.

After basic training, advanced training in munitions, and jump school, I was assigned to the 173rd Airborne. I later requested a transfer to the 335th AHC because flying sounded like more fun than ammo humping. I was a door gunner in the Ramrods and extended my tour twice. I flew in over 400 missions. I was honorably discharged in January 1969.

Shortly after returning home, I met my future wife, Susie, in February 1969. I worked a couple different jobs, finally landing a job driving a truck for a building supply company. I married Susie on July 26, 1969.

I waited to get married one week after my 21st birthday because at that time in Oregon, if you were a male under the age of 21, you had to have your parents' permission to marry. After my stint in Vietnam, I sure as hell didn't think I needed my folks' permission! After getting married and having our son, Michael, on May 2, 1970, I decided to go to school on the G.I. Bill. Since I only had a GED, I attended Portland Community College for one year to play "catch up." After that, we moved to Corvallis, Oregon where I enrolled at Oregon State University. My major was Forest Management. It was really a fun time of our lives, married with a small child. We lived in Adult Student Housing and were surrounded with other young people in the same situation as us. I graduated in June of 1976 and went to work immediately for the Oregon Department of Forestry. My career focused mainly on environmental law and wildland fire. I retired as a Protection Unit Forester after nearly 27 years with the Department.

We spent most of our lives in Oregon but after retirement we moved to Washington state, to be closer to our two grandchildren, Katharyn and Michael. We have enjoyed traveling and seeing the rest of the world, something we only dreamed about.

When asked, "How did your military experience affect your life?" I can honestly say that I entered the Army as a 17-year-old kid and came out a 20-year-old man. Military life made me rely a lot less on my emotions, but I feel blessed to have made it home when so many did not.

DAVIS, WAYNE B.

After leaving the Cowboys in April 1967, I continued serving in our Army until retirement from it in 1983. Upon leaving the Cowboys, I attended the US Navy Test Pilot School at Patuxent River, Maryland and was then assigned as an engineering test pilot at the Army's flight test facility at Edwards AFB, California. While there, I worked on several flight test programs designed to improve the combat capability of our CH47, Cobra, and Y-03A aircraft. After another, but abbreviated, Vietnam assignment, I went to graduate school at Georgia Tech, and was then assigned to the Aviation Systems Division on the Army staff in the Pentagon and worked there for three years. While there, I worked with the group responsible for doing the "leg work" on starting, funding, and defending the Army's top aviation programs (among which were Blackhawk, Apache, Hellfire missile, GPS, night vison, and target acquisition systems). I then returned to ground duty as the Division Engineer of the 5th Mechanized Division. I then returned to Georgia Tech for two years as the head of the Military Science Department. After Tech, I returned to aviation research and development assignments on Army and Department of Defense staffs in the Pentagon.

After retiring from military service, I worked for five years with NORTEL Federal Systems and was responsible for managing the installation and implementation of communication systems at federal government international facilities. After moving during 1988 to Knoxville, Tennessee, I have been actively involved with automobile and power equipment businesses and have owned and operated a John Deere equipment dealership since 1991.

Although I served with the Cowboys some fifty years ago, the Cowboys have never left me. I recall many combat experiences, but what lingers most pronounced are the relationships with my Cowboy Brothers-in Arms: our team work, our desire to get the job done, and our commitment to taking care of each other.

DIBBLE, JOHN CLEMENTS

I have procrastinated mightily responding to Dom's reminders but have caved in so those that have not been a stellar success post-Vietnam will know they are not the only one.

I was a Special Forces medic, I intended to go to medical school after my service. I was OJT to Fort McPherson in the ER when DA tasked the fort for EM to take the OCS or Flight School tests. All TDY persons were sent. I took the flight school test. Six weeks later, I was on my way to Fort Wolters. On arrival in Vietnam ten months later, I requested assignment to a medivac unit. Two days later, I was in Dak To. A COWBOY! I am proud to have been a COWBOY and I am blessed to have served with some of the finest EM, WO, and RLO in the military. GOD bless all.

I was offered to return to Vietnam as a RW platoon leader and said no thanks and left the service. For almost a year, I looked for a job but kept getting the same line. You don't have any experience. All you Vietnam guys are drug users. Finally, I gave up and went back to school. Over a period of 14 years I earned my AA, a dual BS in Psychology and Physiology, and an MPA/MBA. I worked my way through these as a low and mid-level federal government minion. I am not proud of my 23 jobs in 22 cities, but will say I have had lots of experience and usually got mad at a boss and quit but they beat me to it and fired me a few times.

My domestic history mirrors my work history and prefer not to dwell on it except to say I have two great sons that became that way without me; much to their and their mother's credit (different mothers, if it isn't obvious). Some things I just can't bring myself to talk about in a public forum. Suffice it to say I get an F minus and its well deserved. I have been near suicide over my failures. The Dale Carnegie Course saved my life.

Vietnam is a mixed bag. Only those that have been in these situations understand what comradery really is. The guys in front can't do it without the guys in back. The guys in back don't get a vote when the guys in front do something stupid. The guys back at the ranch keep the whole damned thing glued together. They seldom get the credit they deserve, but the COWBOYS were a hell of a team. That said, you can only get the shit scared out of you so many times before your brain becomes rewired. Each of us has a different set of experiences and each has had to deal with it effectively alone.

I hate Lyndon Johnson and Walter Cronkite with a passion. I don't trust the government at all. I hate it more now than I did then. I have been back to Vietnam four times now, I was there last year to celebrate my entry 50 years ago and will be there August 2018 to celebrate my departure. That country is still devastated. Agent Orange is still killing people and their crops are still stunted; and these are the people we were defending! Helping?

The only bright spot is the people are staunch capitalists because millions died when their communist government couldn't feed them. The communists are still in power but look the other way while the people thrive in an overt capitalistic fashion. The communists do keep order and when the lights change, no one runs them. The streets are clean and orderly. The people take responsibility for feeding themselves. I hate communists and Marxists and their socialist wanna-bes.

I live in Southeast Asia now. I have a house in the Philippines and travel from there. I watch the freak show in DC and fly back to vote. I was blessed to be in the States for the dedication of the monument at Mother Rucker and the ceremony in remembrance of Hill 875 fifty years ago which was held at Yanks Air Museum in Chino, CA.

DIXON, JON CHRISTOPHER

I was a First Platoon (Ramrods) crewchief in the Cowboys, July 1969 to July 1970. My assigned aircraft were 076 (Charlie Charlie), 633 (a real veteran, she was) and the exceptional 384, all H-model slicks. I was first to fly Nitehawk missions during the second half of my tour.

I had volunteered to go to Fort Rucker to teach in the Crewchief school after my tour…but of course by that time the need was winding down. After Vietnam, I was stationed in Fort Hood, Texas in a self-propelled artillery battalion. I spent 1.5 years as a motor pool clerk! The battalion finally changed to an aviation (helicopter) battalion, but we never saw even one helicopter. I was discharged in November 1971.

I had a lot of back pain in my adult life, but thought it was muscular. I was not until 2007 that I learned I had broken three spinal vertebrae in a "hard landing" while in Vietnam. Apparently, the surgeon said, youth, adrenaline and the circumferential type of fractures were not problematic until they started to calcify badly in my older years… then the pain was excruciating. Two surgeries down, maybe one more to go.

After leaving the service, I worked at various jobs, mostly managing restaurants in the Cleveland area, where I am originally from.

My fiancé and I were married in 1972 and our daughter was born in 1975. Treasures, both!

My wife asked me once after a long night at the restaurant, "What is it you really want to do?" I immediately replied, "I want to be an aerospace engineer." Once the words were out, I was surprised at what

I had said. I had no idea where they came from. I had never considered engineering at all, let alone aerospace engineering. In retrospect, it was the amazing scope of aviation I saw in Vietnam that must have captured my imagination more than I had realized.

I went to school at the University of Minnesota (where we had moved by then), starting at night, transferring to days and working through a great internship. I graduated with my Bachelor of Aerospace Engineering in 1986 and went to work for a Minnesota aerospace company called Rosemount. A fabulous place to work!

I later earned my Master's in Manufacturing Systems Engineering, and later a Doctorate in Educational Leadership. My doctoral dissertation drew parallels between corporate acquisitions (large companies acquiring smaller ones) and imperialism. (My dissertation's dedication was to "…those Cowboys whose last moments in Vietnam, were their last moments." I have been an aerospace engineer my whole career. I also taught courses in "Excellence in Design" at the masters and doctoral levels at the University of St Thomas, in St Paul, MN for 20 years.

Rosemount was acquired by Goodrich in 1993, and Goodrich by United Technologies in 2012. I have stayed at the same location through both acquisitions. I have worked on military programs (some classified), a space project, and several commercial airliner and business jet programs. I have experience in manufacturing engineering, research & development, technology transfers, implementation of new technologies, and as a quality engineer. I have been fortunate enough to bring significant new technologies into the world of aerospace. I am still working but getting tired.

Also, I am working (near done) on a book about modern American management practices.

I did not like the military much when I was in it…but it truly made me a better, stronger, more independent person. It made me realize what was and what was not important. The important things can be counted on the fingers of one hand. I would not, in retrospect, trade my military experience for anything.

I still marvel at how well the Cowboys got things done, and what a great leader Major Stiles was. Vietnam taught me a lot, matured me well beyond my years, gave me self-confidence.

I enlisted because I had always (from childhood on) felt I owed a debt to my country, and especially to those countless women and men who created this nation by fighting and beating the most powerful army and navy in the world twice in the span of 36 years. I owed a debt to all those who kept that new, great, free nation independent and prosperous ever since.

I am still a staunch patriot, and fear our country is falling to pieces in a dramatic way. Politicians today think it is their job to tear down the other party, not to work for you and me as Americans. The net result is that nothing gets done to make America a better place. Political correctness has replaced honest opinion. Pride in America seems to be something that is best unsaid in many circles…not mine, but in many.

Dr. Jon C. Dixon

DUMOND, GARY ROLAND

I came back to Maine and went to work for an air taxi business in northern Maine. In September of 1971, I was hired by the Maine Forest Service as a ranger pilot. October of 1972, I transferred to the Maine Warden Service, the law enforcement branch of the Maine Department of Inland Fisheries & Wildlife. Retired from there in 1991. Flew Cessna 185 on floats and wheel/skis. Also flew DHC-2 Beavers, Piper PA-18 Super Cubs, Bell 47-G5s & JetRangers.

Purchased a recreational motorsports dealership (Fort Kent Powersports) in 1990. We handled snowmobiles, ATVs, personal watercraft, outboards, boats, and motorcycles.

After selling the dealership in 2005, I spent seven summers flying Beavers at a few fishing lodges in the Bristol Bay area of Alaska.

I have been married to Pauline since 1971. We have two daughters, one grandson and one granddaughter. We continue to live on Eagle Lake in northern Maine.

Some of my hobbies are snowmobiling, radio control aircraft (big stuff!), fly fishing, and boating.

I belong to Seaplane Pilots Association, VHPA lifetime member, Lions Club, and Quiet Birdman. In November of 2017, I was presented with the FAA Wright Brothers Master Pilot Award. I have logged just over 21,000 flight hours, 11,000 in floatplanes.

I feel like it was an honor to have been assigned to the 335th AHC, Cowboys! Looking back, it is hard to believe so much trust was put into the hands of a 20-year-old AC. Lifetime memories and friends!

Gary (Frenchy) Dumond COWBOY 24 January 1970 / January 1971.

DUPUIS, DENNIS

Cowboy 21

I was 20 years old when I left the Cowboys at Bear Cat in August 1970. I took a 30 day leave before reporting to Hunter AAF, Savannah GA as a class counselor for VNAF students. I leased an apartment with two flight school classmates. We won a one-year rent-free lease contest. I drove a 1970 GTO and bought a 1969 Triumph Bonneville. I asked for an AH-1 transition and returned to Vietnam in August 1971 to fly with the Banshees, B Trp, 2/17 Cav, 101st DIV at Camp Eagle. Our primary mission was recon from the A Shau Valley north to the Khe Sanh plateau. Halfway thru the tour, I received RIF orders (believe or not they are written on PINK paper). Honorably discharged March 2, 1972 at Fort Gordon, GA.

After leaving the service, I worked a variety of part time jobs while attending Augusta College 1972-73 and dropped out.

I struggled to find full-time employment with the stigmatism of a Vietnam veteran. I was proud of my service and continued to fill in the Veteran part of job applications. I met and fell in love with Kay a few weeks after coming home. We were married in 1974. I worked for Club Car (golf car manufacture) as a Service and recall rep. I missed flying and joined the South Carolina NG for weekend duty in 1975. Dick Fogel and Bill Freeman were in the SCARNG.

In 1977, five years after being part of the RIF / lay off from the Army (sounds bitter—I was), the local union of the International Brotherhood of Electrical Workers was proud to have me become a

51

member even though I was a Vietnam Veteran. It was a good feeling. I considered that moment as my official Welcome Home. I owe as much to the Union as I do to the Cowboys. I retired from the Local in 2006. In 2003, I retired (W5) from the SC Guard's 51st Aviation Group as an Apache SIP.

In retrospect, if it had not been for the RIF, I would have never been blessed with Kay. We have two daughters, Amy and April. A thought—why are most aviators blessed with daughters? Anyway—our daughters are married to good men and we have three grandchildren. They must be the reason that I survived two tours. We also have a beagle, who is treated as a child, named Fred.

After retiring, I have been so busy that I don't know how I ever worked 40 hours a week. We often travel 70 miles to the grandchildren's activities or to the numerous military bases to visit the other daughter before John retired, have been on the top of the TO DO list; or to Washington state to see my brother. (1968-69 Bruce was a MACV advisor at Kontum. He was wounded twice and recovered in country. Anne was his nurse. They were married when they returned to the world and are still married. She loved him so much that I think she may have shot him a second time, so she could take care of him). I still ride a motorcycle, it is my wind therapy. I enjoy riding with the Patriot Guard Riders, American Legion riders, and a few close friends. I have been writing a few of my memories and have had four stories published as a chapter in different books about Vietnam. I have spoken at a Vietnam Veterans parkway dedication to help recognize our vets. Google Search "Dennis DuPuis speech." I have participated in a few first-person history panels at local grammar schools, museums, and colleges.

The most notable event in the past year has been the return of a cousin who had been missing for seventy-five years. Cousin Peter W Atkinson flew with the AVG Flying Tigers. He was my inspiration to become an aviator. His remains were located and returned to his hometown of Martinsburg, WV in April 2017 where he was laid to rest next to his family. His brothers, Robert and Al, were also pilots during World War II.

I flew with three different units in country. I have no doubt that The Cowboys and Falcons were, by far, the best crews and the best unit. I include all the Cowboys who supported the flight crews, for without them our Hueys would have been paperweights instead of flyable aircraft.

Here is a little toast that I have for all the Cowboys:

To Those We Knew With Whom We Flew
To Those Here Now And Those Who Are Gone
We Raise Our Glass High
And Say Farewell But Never Goodbye.

Thank you all. God Bless You.

EGET, EDWARD G.

Submitted by: Corinne T Eget

Ed was shot down in Vietnam June 14, 1969 and received the Purple Heart and Bronze Star. He also earned Air Medals with 16 oak leaf clusters and a Vietnamese Cross of Gallantry with Palm while in the service.

His hospital stays involved eleven months of traction and extended to two years with many operations. He also had to deal with the staph infection called osteomyelitis, which caused great pain at times even into his sixties.

However, my husband was a very hard worker and a great family man with five children. After leaving Walson Army Hospital located at Fort Dix, NJ, he went on to graduate from Rider University Magnum Cum Laude with a grade point average over 3.5. While pursuing his college degree, Ed also started a flying club which eventually had twelve airplanes and received a job selling sewing machines. As manager of the Singer sewing machine store, in one year's time he took it from next to the worse store (out of 125) to be the best. Ed eventually bought and ran two sewing machine stores of his own. He also increased his flying club to 200 members and 30 airplanes. In addition, he ran a substantial maintenance charter operation and flight school. He and partners also bought the airport and adjoining golf course.

The long hours of work and the stress of an aviation business took its toll on Ed and he had to have open heart surgery in 1986 and a 10-hour operation on his leg in 1987.

In 1993, we left the Aviation business and built a world class golf course designed by Gary Player. We opened the course in 1995 and were recognized by Golf Digest as one of the best courses in NJ.

Still working hard to make the Old York Country Club a success, Ed had another heart attack in 1999. It seemed his first open heart surgery was projected to last 15 years. After his heart attack, Ed cut back on his work hours. In 2006, we were able to have him accepted at the Cleveland Clinic in Cleveland for a second open heart surgery, by the head of the Coronary Department. The operation was a remarkable success, but unfortunately, Ed died November 8, 2010 of large and small cell lung cancer.

Ed had a difficult life but always worked hard and stayed positive. He always made those around him feel good about themselves. People loved to be with him because he was so much fun. He was a wonderful person. I could not have imagined a better husband, father, stepfather, or friend.

EGGERS, WILLIAM HENRY

When I left Vietnam after my "365 Tour" I flew into Oakland Replacement Station in California. The steak we had that first night was great! I never lost sight of that steak house that year. I was coming back and was going to love it!

I flew into JFK Airport a few days later to my parent's home in Garden City, NY, then a short train ride from the airport. I either didn't have any cash on me or I just wanted to surprise my parents by walking in and saying—I'm back! It is noted that my flight to NY was a lonely one. As the drill sergeants told us in boot camp, "Don't expect a hero's welcome when you come back home from Vietnam." They were right—no one said a word to me. Even though I was in uniform, no one in the airport, the plane, or the train ride to Garden City said a word, or even smiled at me. I remember well that this occurred. I was neither surprised nor angry, that's what I expected and that's what I got. Anyway, my parents were surprised, and it was great to be home.

When I showed up at the US Army recruiter station back in September 1970 (I had a low draft number—#113) I was feeling real fine after bar-hopping with my older sister earlier in the day. Long story short, I had just graduated college in May and, not being in ROTC and not having well-to-do parents that had connections, I went to all the services and attempted to go into Officers Candidate School. I was told by all the services that there was an 18-month waiting list! So, I entered the Army as a private.

The Army recruiter told me that I should enlist as an Airborne recruit, otherwise the Army will just might make me a grunt, a cook, or whatever slots they needed filled. As said, I was feeling somewhat drunk that night and told them I might be drunk, but I know enough not to jump out of any kind of plane! So, I ended up signing up to be an MP—it was guaranteed but it cost me another year of enlistment (3 years).

So, after I returned home from Vietnam, I still had 18 more months left. I was stationed at the US Military Academy, West Point, for the next 18 months. I was a Patrol Supervisor (Sgt.) at West Point. You might be asking yourself how was it that I was assigned to the 335th Assault Helicopter Company in Dong Tam for six months? Long story short, my buddies that I was with in Fort Dix boot camp and Fort Gordon, GA, MP School all made a pact that we would not do any drugs in Vietnam. The drill sergeants drilled that into our brains every day—"Don't even try any of that 'skag' (heroin)." Well, they all pretty much, with the exclusion of me and two others, broke their word and got hooked on drugs. The peer pressure was immense—I had better things to do in my life and didn't want to get mixed up on drugs. I never even tried the skag that was put in front of me endless times. One night in Long Binh, where I was stationed, I walked out of the NCO Club and got robbed of my wallet, not before being hit in the face with a piece of 2 x 4. The drug scene was bad. I asked 'Top' if there was another more professional unit to transfer to. He totally understood my motive and in a few days, I had orders to travel to the 335th AHC in Dong Tam. The rest is history.

So, back to West Point, where life was good! I met a girl at one of our MP barbecues at one of the lake-side beaches. She was attending Ladycliff College across the street from the military college. We got married a year later and lived at West Point until my ETS in September 1973.

At the time of my ETS, my wife's parents were planning to move to Colorado. It made good sense for us to move there and have the Army ship all our belongings to Colorado. A few weeks after ETS, we were residents of Pueblo, Colorado. The police department was not hiring, there was a two-year waiting list. I ended up working at the Holiday Inn across the street from where we were living. The head of the Maintenance department was an Air Force veteran and hired me on the spot. I learned to do all sorts of stuff which later in life came in very handy. Cecil taught me how to solder pipes, weld, and to fix motors and other electrical appliances. To this day, I thank Cecil for teaching me those trades. Within a year, I was looking for a better paying job, one of which was the Colorado/Wyoming Railroad. I was told, like the police department, come back in two years. When I returned home that day, my wife told me the railroad had called and wanted me to call them. I did and was told to report to the employment office the next day. The next day I was hired at a manual labor position on the rip-track-gang, which paid an enormous amount of money! The reason I was hired so quickly was a matter of perfect timing for me. The EEOC was in full swing to ensure that the minorities were treated fairly. Believe it or not, I was deemed to be a minority because the census showed that the population was 54% Spanish! Me and three other "WASPs," as they called us, were hired and started the next day!

Working the rip-track-gang was extremely difficult. I had put on a lot of weight after getting married and slinging a pick and slamming it down on a spike all day was tough. We did not have any machines to do the work. It was all done manually. We ripped damaged track out and put new track in. The AFL-CIO was the employees labor union. That's one reason the pay was so good. However, when told we had to work overtime to get the jobs done, we could not say no. If we did, we would have gotten fired. So, many a night I came home past midnight and had to be back at seven in the morning. Even though it was tough work, the high pay was the reward. After three months on the railroad, the local newspaper had an article saying 22 police officers were fired for participating in a burglary ring of a tire warehouse, thus, leaving 22 open positions. I really did love police work. My mother told me since age eight that I would be a detective one day!

So, after passing the civil service test, I was then scheduled for the physical agility exam. Here was another time when everything fell right into place. The police department had to open sworn police positions to women, according to the EEOC! This was a nation-wide ruling, not just in Pueblo. So, now all the civilian female police department employees could apply for a sworn position, and possibly being promoted to police officer. I was lucky to get some inside information that the physical agility test was going to be "boosted-up" so as to make it more difficult for females to do well on the physical agility exam. The 160-pound body drag was now going to be 180 pounds and some of the other stations made it difficult to max out on the pull-ups, pushups, run-thru-the tires, etc. I would have been shaking in my boots (I wore cowboy boots back then) if I had not been working for the railroad. In three months, that job got me back into shape, just as Army boot camp did. I was in top physical condition. I aced the physical agility exam and was hired by the police department.

On my last day at the railroad, a manager came to me asking if I wanted to stay and be bumped up to engineer. I turned it down. No regrets.

A little over two years with the Pueblo Police Department and with one child and one on the way, my wife told me that she did not care to live in Colorado any longer. Although her parents moved out there, she wanted to move back to the East coast where she grew up. One of the motivators was the fact that one of the newer middle schools in Pueblo started up with the "open concept design" but more importantly, on the outside entry doors was written — "Spanish" and another door — "Whites." La Reza, otherwise known as the Brown Berets, was the gang that plagued the area. My wife said she didn't want to raise our children in an environment like that. At the time, we had a one year old and one on the way. Just before we moved to New

Hampshire in 1986, I delivered a little baby girl in the front seat of a car at midnight on a July evening while on duty. It made front page of the newspaper the next day!

We settled in Hollis, New Hampshire, a small town just west of Nashua. We had taken a vacation while I was at Pueblo Police Department. The Milford Chief of Police said if we moved to New Hampshire, he would hire me. That's what happened a month later. I spent ten and a half years with the police department, being promoted to detective a year after my arrival, then to sergeant, then lieutenant.

At the time, I was on the 'short-list' to attend the FBI Training Academy. While I was a detective, I was also the Juvenile Officer/Prosecutor in juvenile court. That was a blast. I was arresting kids and prosecuting them! New Hampshire statutes allow a trained/certified police officer to become a District Court Prosecutor. Basically, it was a way to allow the cities and towns to have police personnel act as prosecutors rather than paying for city attorneys.

In hindsight, I can clearly say now that my military bearing did have a very large portion of my success in the positions I have held. It didn't seem so back then, but being in the military was where I seemed to have found myself.

Unfortunately, after ten years with the police department, in August 1986 I had a disabling on-duty injury to my left knee which the orthopedic doctors in Boston (Carter Redd Rowe and Bert Zarins) said they were not able to repair the knee. Dr. Zarins, the orthopedic surgeon for the New England Patriots told me even if I was making eight million dollars playing ball, there was no way the knee could be repaired. They did not have the technology to repair a torn posterior cruciate ligament back then. Due to that diagnosis, I was forced to retire on a State of New Hampshire disability pension.

Six months later, in March of 1987, we moved to Cape Coral, Florida where we remain in the same home. By then, Margaret and I had five children. By 1993, we had seven children. Within six months after moving to Florida I became a Licensed Private Investigator. It was a very cool job. I was hired by attorneys to investigate slip and fall allegations, murder investigations, divorce, child custody, etc.

Four years later, 1991, I was hired by the State of Florida Department of Juvenile Justice as a Juvenile Probation Officer. I still wore the brace on my left leg. I was still limping. I was on an upward mobility track with DJJ, being promoted to Detention Review Specialist position for three years and then to a mid-management position as the Education Coordinator for the Southern District of Florida.

After 9/11, the state legislature decided to cut hundreds of state positions as they said the economy will suffer from tourist who would not be flying to Florida! They estimated that the state would lose 25% of tourist dollars in the first year following 9/11. So, as a result, 600 plus mid-managers like myself were handed 30-day pink slips. We got the pink (it really was pink) slip on 12/07/2001 and four other co-workers in the office and I had our going away party January 7!

From there I went back to school! Three months of sending out job applications led me nowhere. I was fortunate to go to the county employment office and was offered to be "retrained." I chose the Microsoft Certified Systems Engineer track, since I always wanted to know more about computers. The school was very difficult. I was the oldest student; the rest were college aged men and women. I sat in the last row (theater style) which was my comfort zone. Before long, as the group became friendlier with each other, the boys were hitting on the girls all day, which became a big distraction for me. This was heavy stuff to learn. I was sitting up higher than the rows in front of me and was seeing everyone's monitors. I told Margaret I had to move my seat because of all the "grab-assing" going on! This was a $21,000 school that these people paid for. I was lucky. Because I was a "displaced state employee" I did not pay anything for the class. It was a great learning experience, much more difficult than college courses. Although I was the oldest person in the class, I was the first person to take the exams.

It was "at your own pace" system. You could only take a pencil and one piece of paper in the testing room. We were even searched before we went in! It was very rewarding to know I was the first one to do it—and to pass it on the first attempt.

I was collecting my State of New Hampshire police pension and unemployment at the time. I was still job hunting but wanted to obtain more knowledge about computers. By early 2003, I was ready to go back to work. I was hired back by the State of Florida at the Guardian ad Litem Program (GAL) in a supervisory position. The GAL job was a very demanding position. We advocated for the best interests of children who had been abused, abandoned, or neglected. These kids were usually placed in foster homes while their cases progressed.

Our oldest child, Daniel, was a Special Forces US Army Captain (Green Beret) stationed at Fort Bragg, NC. He was married with two small boys. While on his second tour of Afghanistan he was killed in action in 2004. It took a great toll on our family. Life took a different turn on that day—at least for me. I was the veteran warrior who thought nothing could bring him down. I was wrong. When one of your children dies, it becomes a life-changer. I had to be strong for the rest of the family. By 2015, at the age of 67, it felt like I was burning out. I needed a rest from the depressing thoughts of all the broken and dysfunctional families I had been dealing with for the past twelve and a half years at the GAL Program.

On September 4, 2015, I retired from the State of Florida with 23 ½ years under my belt. One the first things on my bucket list was to sleep late in the morning. That never happened. Even to this day, I am up and out of bed before 7:00 am!

The second item on my bucket list was to build a memorial for my son at the Cape Coral Military Museum. That was done, and it was completed on December 14, just five minutes before Jessica Lynch walked into the museum. Jessica is the famous soldier who was captured in Iraq where she was tortured and kept in deplorable conditions. She was eventually rescued by US Special Forces. Jessica Lynch was visiting the museum that day—and coincidentally I had just put in the last screw as she walked in! When she saw the exhibit, she said she was blown away—then she wept.

The memorial I built is complete with a TV that shows videos and slide shows of photographs CPT Eggers took on his first tour of Afghanistan as well as the original plaque from Camp Eggers, in Kabul, Afghanistan, which was named in honor of CPT Daniel W. Eggers in 2005!

If nothing else, being a member of the United States Army showed me the way to become a man. Even though I was a college graduate going into the Army, I had a lot to learn. While I was in Vietnam, I won the prestigious Platinum Tail Rotor Chain Bracelet, a result of being "shot down" twice. The first time the enemy took a piece of tail rotor off and we auto rotated to a smooth landing. The second time a bullet came through the hell hole and hit one of the hydraulic lines. After a CA mission, we were headed home, the last one in our group. As I poked my head out and looked behind us, I saw a trail of black smoke as far as the eye could see—coming from us!

As soon as I announced our problem to the crew, the bells and whistles went off. Another auto rotation, not so smooth that time. We had no hydraulics. The third time, I had time to speak to God. I made some promises which I have kept to this day. We were on a mission, not a CA, but like a mail run, except we were in a hostile area, we couldn't land. A buzzing sound developed in the nose of the aircraft, which soon developed into a grinding sound that reverberated throughout the whole deck of the helicopter. The vibrations ran right up your spine. When we successfully got back to Dong Tam we were told that the 'inverter' had gone bad and was making that racket. I became a true believer that day and have been since. I knew God much better by the time I returned home alive.

Some of the organizations I have been affiliated with are the American Legion and the Veteran of Foreign Wars.

Our three boys were members of the Young Marines of Lee County (Florida) in the mid 1980s through the early 1990s. The Young Marines is the youth organization of the Marine Corps League. Their headquarters in Michigan at the time appointed me as the Acting CO. No other dad would step up to the plate. They said I was the one and only non-Marine to hold that title. We took the kids to Camp Lejeune one summer and to Parris Island a year or so later. We had a very tight unit. We won lots of honors and awards. In 1993, my son Daniel was named Young Marine of the Year (Florida). A few weeks later, he accepted the honor of being the National Young Marine of the Year.

I have been the President of the Men's Fellowship Ministry at St. Andrew Catholic Church in Cape Coral, Florida for the past twelve years.

Although I am not a paid staff member of the Guardian ad Litem Program any more, I have been a GAL Volunteer since I retired.

I am retired; however, I have submitted many employment applications here and there. Many have been with the State of Florida. I've had some interviews, one of which I am awaiting on whether I'll be hired or not. Eating and living healthy has given me a wonderful outlook on life.

I feel I've still got a lot of energy and enjoy the working environment.

I am lucky enough to not be on any medications, other than vitamins and fish oil! I have taken advantage of being a patient at the VA Clinic in Cape Coral, Florida. They are great people and they roll out the red carpet when you walk in. Other than getting an annual physical each year, I am going to try out their Ophthalmology department this week for a new set of eye glasses.

Margaret and I have seven children and eleven grandchildren.

Thank you for the idea of writing our past experiences following Vietnam. I actually got into this task and will pass this to my children and grandchildren. Sorry for the long litany, but I had the time to share part of my life with you all!

ELLSWORTH. MICHAEL L.

I joined the Army in 1969, and went to basic training at Fort Bliss Texas, and AIT at Fort Rucker Alabama. After AIT I received a 30 day leave and spent that time with good friends in Spokane until I departed for Vietnam.

After Arriving in Vietnam, I was eventually assigned to the 335th AHC in Camp Bear Cat.

Later that year, we had the big move to Dong Tam. I served as a crewchief in the 1st platoon, until I left in the summer of 1971.

After Vietnam, I went to Bad Hersfeld, Germany. I was a Kiowa crewchief on a brand new OH-58A. It had 25 hours on it when I was assigned. Our duty was to fly border patrol on the Czechoslovakian border. I left Bad Hersfeld and the Army in December 1971 and moved back to Spokane.

I started work building boats at Fiberform and going to school full time at Spokane Community College, where I earned my A&P license.

It was about this time, in 1974, that I met the woman that I married and had one son. We are still married and are currently living in Las Vegas.

I hung around Spokane until 1989 working in a grocery warehouse, until I decided to utilize my A&P license. I worked in Alaska as an aircraft mechanic until about 1995 and moved back to Spokane and worked for a DC-3 operator until I moved to Las Vegas and accepted a job as the director of maintenance for a Grand Canyon tour company.

I was hired by Hawaiian Airlines in 2002 as a maintenance supervisor and have been with them ever since. I plan on retiring this year.

I started my aviation career with the Army and built on it after the war.

All in all, the Military provided me with an illustrious career field that I still work in today.

FINEGAN, RAY D.

1967—After returning from Vietnam, I was assigned to Hunter Army Airfield as a contact IP and Flt Commander, great duty. Got married and the phone call for the second tour, went to Cobra transition and Aviation Maintenance Officer Course. During this period, I survived a private plane crash, lost an engine on take-off from Savannah at night. Those pine trees are sure tall, but provide a great cushion.

Second tour I was assigned to the Americal Division, 14th Aviation Bn as Battalion Maintenance Officer. This was shortly after Lt. Calley had done his thing, did a lot of firing incident investigations, did not enjoy going to the villages.

After Vietnam (1971), I went to the TC Advance course and then back to HAAF as the CO of the Cobra Maintenance Company, a very rewarding experience. My wife had a baby boy (Mark). After command I was Bn XO and received that call once again, but this time we were going to Iran as part of a Technical Assistance Field Team for two years. Very interesting, many of the officers I worked with were later executed.

1973—Returned from Iran and went to the Air Command and Staff School at Maxwell AFB, there were 25 Army officers and it was a great year vacation before getting back in the real world. Went to TRADOC and then to Korea (KMAG)

1981—After Korea, I was assigned to AVSCOM and TROSCOM in St. Louis as Deputy Commander of the Field Service Activity. DA made me an offer that I could refuse, so I retired in October 1986.

After retirement, I couldn't decide what I wanted to do when I grew up, so I went into residential real estate. My wife at that time decided I was no longer worthy of her company so we divorced. By this time, Mark was at college in Wisconsin.

In 1995, I married my manager, Lynn Bodenheimer. This year will mark Lynn's 43rd year in real estate and my 31st year. We still work as a team. Lynn and I enjoy the theater and traveling. We spend long weekends in Lido Beach (Sarasota, FL) and three or four weeks in Maui. Our travels have taken us to China, Hong Kong, North and South Vietnam, Thailand, Singapore, Japan, Alaska, Greece, Croatia, and Italy. Planning our next trip.

Life is good, I am a cancer survivor and have been cancer free for three years. EVERY DAY IS A GOOD DAY, SOME ARE BETTER THAN OTHERS, but it sure beats pushing up daisies.

The 335th holds many fond memories and was my favorite assignment in Vietnam because of the leadership and comradeship.

FINO JR., DOMINIC PAUL

After graduating high school, I went into the Army. I was 17, young and dumb. I went to Vietnam at age 18 and was married before departing for my first tour. On my second tour, my then wife and I had one daughter. My daughter now has a daughter of her own. My granddaughter now has produced a great granddaughter.

I left the service in 1971 after three years in, at age 20, and could not vote or legally buy a beer. Most of the Country did not like me, as I was classified a Vietnam Veteran. I have never forgiven the war protestors for their behavior, nor have I forgiven the government for their incompetent management of the war.

The time I spent in Vietnam taught me all I needed to know about teamwork, process improvement, and doing a job to the best of my ability. As a person, I grew up in Vietnam and have never regretted a single day I was there. Realistically, flying as a crewchief on a gunship was the most important, dangerous, exhilarating, and rewarding job I ever had.

After the service, I went to work for the Baltimore Gas and Electric Company as a cable installer. Apparently, Vietnam Veteran helicopter crewchiefs were not in high demand. There was even a lesser demand for folks proficient with an M-60. Go figure!

I immediately enrolled at The Johns Hopkins University evening college under the G.I. Bill and attended classes three nights a week for seven years while working full-time in various positions at the utility. I graduated from Hopkins with a Bachelor of Science Degree and went on to become Supervisor of Employment at the utility company.

After twenty-three years at the utility company, I was caught up in the downsizing movement and released with a year's pay. That is when I learned the lesson that corporations do not have hearts, only bottom lines. I did manage to finish a non-fiction book about the Falcons during my downtime. It only took 20-years to complete, but I then had more time than money.

In 1992, Doug Wilson contacted me out of the blue and we decided to get the Falcons back together again. The effort was soon expanded to include all the 335th and Susan Posey joined us in our efforts to find as many former Cowboys as possible. As they say, the rest is history.

What happened next was not good. My wife of 23-years divorced me, I had just lost my job at the utility, and my favorite dog had to be put down. Yes, it is true. I still miss my dog!

Life was not looking too good at this point, but things change. I opened a packaging and shipping business in Baltimore County and ran it for two years. It was taking more time out of my life than I wanted to dedicate to running my own business. The hours are long and there is never time to fully relax. I then sold the business and went into full-time consulting work as a project manager for several companies, developing and installing software and networks. Skills I picked up on my own working alongside some very smart people. This was the best avenue for me since I could control who I would work for and for how long.

I remarried and managed to write two more books. Both are fiction murder mysteries. My current wife is very loving and understanding when it comes to the time I put into the 335th Association. She understands that it is not only important to me, but to all our members.

I have never been a joiner. That is, I do not belong to any organized religion, I have no allegiance to any political party (registered independent), and I do not belong to any other groups or organization except for the 335th AHC. I also have developed a very strong dislike for most politicians and lawyers. They are prostitutes. The only difference is prostitutes will stop screwing you.

I fully retired a few years ago and do not miss working. I never could fully accept the idea of having others tell me what to do, how to think, or how to live. Besides, I have mastered the art of screwing up and then fixing my life without others interjecting their ideas or views into the equation.

In conclusion, life is good, and my health is mostly good. My two goals now are simple, to keep the Cowboy spirit alive and to outlive my enemies.

FRASER, DAVID LANCELOT

1967 – 1969

I entered the US Army at age nineteen. My family had emigrated from Scotland when I was a boy just shy of nine. My path to US citizenship lay with the military — serving in combat would provide immediate eligibility for citizenship. I enlisted via the Army's *Buddy* program with my high school friend and fellow Civil Air Patrol cadet, Johnny Cook. Together, we took our Army basic training at Fort Ord, California, in March/April 1966. Meningitis was a constant threat on base and we were prohibited from closing the barracks windows — it was cold and damp. I was hospitalized twice during basic due to complications of asthma and was offered a medical discharge. I declined, telling the doctor I needed the Army as much as the Army needed cannon fodder. I was allowed to remain in the Army and subsequently graduated basic training with my buddy, Johnny Cook.

We both went on to Airframe Maintenance school at Fort Rucker, Alabama. I then continued to Crewchief school for UH-1, *Huey*, and then to CH-47, *Chinook*. In the meantime, my application for Helicopter Flight School, Warrant Officer Candidate, was accepted. Johnny Cook had also been accepted to the same Flight School class — once again we were in training together and although I was set back a class due to problems with learning to autorotate, we both graduated from Flight School and earned our Army Aviator Wings and Warrant Officer bars. We then received orders for our next separate postings — Vietnam.

I arrived in-country after Thanksgiving, 1967, and was subsequent-

ly assigned to the 335th Assault Helicopter Company, *Cowboys*, at Phu
Hiep, II Corps, Central Highlands. I joined the 1st Platoon, *Ramrods*.
On my first in-country *Huey* check-ride with my platoon IP (Instructor Pilot), we flew to the Kontum/Dak To area to assist with follow-up
action from the recently concluded Battle for Dak To, Hill 875, involving the 173rd Airborne Brigade. Those first experiences of the consequences of combat have remained with me to this day.

Later, I was temporarily assigned to Kontum airfield, adjacent to
the 57th AHC, Gladiators, compound. While attending the Gladiators' chow line, I saw a familiar face—my friend Johnny Cook was
flying with the 57th. We had a good reunion and were able to spend
several days regaling each other on our activities following each day's
mission completion. On January 10, 1968, the 57th was attacked and
overran by Viet Cong sappers. For an undetermined reason, the VC
bypassed our two minimally protected 335th Huey helicopters and
concentrated on attacking the 57th compound. Several guys from the
57th died that night. Why Charlie passed us by I'll never know—by
the grace of God, or fate, we weren't their target that night.

A few weeks later, Johnny's 57th Huey helicopter was hit by a
North Vietnamese RPG (rocket propelled grenade) while on an in-
sertion mission in Laos. Johnny died of wounds on February 29, 1968.
Warrant Officer John William Wayne Cook is buried at Arlington
National Cemetery. For me, the rest of the war was a matter of surviv-
al and revenge. Other than random flash-backs of events, I don't really
remember much.

I eventually achieved AC (Aircraft Commander), flying insertions,
extractions, re-supply, ash and trash, Command and Control, etc. In
September 1968, I was re-assigned to the 268th Combat Aviation Battalion, Black Lightning, also located at Phu Hiep, were I worked Battalion Operations, Mission Control Center. As my tour of duty came
to an end, I decided to extend and use my thirty-day leave to return to
the States and gain my US citizenship.

Although I had applied well in advance, the INS would have none of my rushed application for citizenship. For three weeks, I visited the Los Angeles INS building and was repeatedly told, "Come back next week!" On the fourth week, I dressed in full Class A uniform, Chief Warrant Officer bars, aviator wings, and combat decorations. When told again to, "Come back next week!" I refused to leave. I was scheduled to return to Vietnam for my second combat tour and I demanded my rightful US citizenship. I was prepared to go AWOL if needed—at least, that's what I told them. Toward the end of the day, I was shown to the INS Regional Director who had arranged a three-way telephone conference with a Federal Magistrate in Washington, DC, and the INS office in Atlanta, Georgia, where my initial application for citizenship had been located. I was sworn in as a Naturalized Citizen of the United States via telephone conference. My Oath of Citizenship was signed, witnessed, and fulfilled on February 5, 1969. I then returned to Vietnam to complete my second combat tour—this time as a United States citizen.

My post Vietnam assignment was to Fort Eustis, Virginia (intimately know as *Fort Useless*.) The Army offered me an early out in May 1970. I had become engaged to marry the first woman I met following my return from Vietnam. She wanted nothing to do with the military—it was an easy decision. I received an Honorable Discharge and was married within a few weeks. My decorations include the Bronze Star Medal, Air Medal with 23 oak leaf clusters, Army Commendation Medal, Vietnam Service and Campaign Medals, etc.

Because of my combat flying experience, I was accepted to join the Los Angeles County Sheriff's Department to fly their new TH-55 helicopters—the same in which I received my primary helicopter training. I graduated from the Sheriff's academy as a Deputy Sheriff, only to learn that a political shift in Sheriff's policy precluded me from their Aero Bureau. I tried working the county jail, patrol, and eventually ended up driving the big passenger busses taking inmates to and from court. I gained an interest in court procedure and the nuances of criminal law—it was to change my career direction.

My personal life was not going well. I had returned from Vietnam an angry young man—and was then given a badge and a gun. I became a person I didn't like. My marriage failed. I went to the VA; but there was no help there. I sought private psychological counseling and slowly started to realize the unpleasant person I had become and took the steps needed to change and be at peace with myself.

While working the court system, I learned of the Public Defender's Office and subsequently left the Sheriff's Department and started a new career as a Criminal Defense Investigator with the Los Angeles County Public Defender. I retired thirty-five years later as a Supervising Investigator—I thoroughly enjoyed the work.

Along with my change of career, I found peace in the solitude of the San Jacinto Mountains of Southern California. I bought a small cabin in Pine Cove, CA, and was able to rest at the end of each day, although my work AO (Area of Operation) was a hundred miles from where I had chosen to live. I needed the separation that mountain solitude provided from the criminal chaos of the Los Angeles city and county environs.

I returned to my Scots origins, competed at local Scottish Games, and revived the kindred spirit I'd known as a boy in Callander, Scotland (known as the gateway to the Scottish Highlands). I returned to college, rekindled my Fraser heritage, and met a Scottish lassie who put up with me and enjoyed the mountains as I did. We had the same surname of Fraser and our children have grown up in the mountains as local Fraser-Fraser *Hillbillies*.

I retired at age sixty-three, following almost forty-four years as soldier, deputy sheriff, and criminal defense investigator. I still have flashbacks of Vietnam events—the smells of blood and cordite are never really forgotten. Some of my habits are unusual to those who haven't experienced combat. The VA denied my claims of Agent Orange caused chloracne and cancer of the neck. Following years of appeal, denial and appeal, I was eventually afforded a disability for PTSD. In retirement, I've been active at our local American Legion Post. As both Post Adjutant and Post Service Officer, I've been able to assist others who share the same daily PTSD afflictions I experience.

My wife (now approaching 40 years of much patience on her part) and children (an adopted daughter and a son from my first marriage and a daughter and son from my second marriage) have come to recognize that combat has lifetime consequences and that they, too, share those consequences. There's not a day that goes by that some sound, smell, or event brings to memory a thought or image of those times of combat fifty years ago. It stays with me—always just over my shoulder and never quite forgotten or out of mind. A sudden move; a loud bang; crowds; are but a few of the frequent triggers.

This immediate past Veterans' Day, I proudly wore my kilt and military decorations. I spoke during our Post's ceremony in honor of the many immigrants who have served, and still serve, in our country's armed forces and those who have, and are, engaged in combat action. A sizable percentage go on to gain their US citizenship.

It's a good thing!

FROMHARTZ, JOHN JOSEPH

I returned from Vietnam in June 1966. I was discharged from the Army three days later from Fort Dix, NJ. I boarded a plane and flew to LA, where I was planning on going to school. I met up with Lee Hall, a fellow trooper who returned six months prior. We decided to jump in his car and travel the States. Two months later, we arrived in my home town, Fort Lauderdale, FL.

I kicked around with several different jobs. I was a fire fighter for about a year. Then a police officer for a while. I found these both to be a little too military. In 1968, Bendix moved their Avionics division from New Jersey to Fort Lauderdale. Avionics is what I did in the Army. I went to work for Bendix in late 1968.

I met my future wife in 1970. She was a manager of a large Mexican restaurant. I also started college the same year. I left Bendix in 1971. My girlfriend trained me to be a bartender. I tended bar for the next 30 years, mostly in popular restaurants in Fort Lauderdale. I ended my bartending career as the beverage manager of the Fort Lauderdale Country Club.

I married my wife Carol in 1974. The same year, I graduated from FAU, Florida Atlantic University. We had our first child, a girl, in 1976. The next was a boy in 1978. In 1976, my wife decided to buy into the Mexican restaurant. She owned and operated the restaurant for the next twenty years.

At 56 years old, bartending became a little old. I've always been a boater. I have a 25-foot center console fishing boat that I use to take me to the Bahamas. Fifteen years ago, I went to school to obtain my captain's license. The day I received my license, I landed a job with a local water taxi company. I retired from the water taxi twelve years later.

My wife Carol passed away in 2014. She was legally in charge of my granddaughter, Wynter. Because my wife had passed, I was put into the position of adopting my granddaughter who was 14 years old. Not realizing how expensive teenage girls are, I'm back working at the water taxi. She's in her first year of college at FAU.

I am a lifetime member of the DAV and I belong to the American Legion post 142 here in Pompano Beach.

GAMMONS, VANCE S.

I was rated in August 1959, followed by tours in Korea, Fort Rucker, France, and Germany.

I arrived in Vietnam in January 1964 and spent the year in the II Corps area with the 52d Avn Bn's two companies, the 117th and 119th AHC, at Qui Nhon and Pleiku respectively.

Following the aviation safety course in U S C, pathfinder course, and the infantry advanced course at Fort Benning, I returned to Vietnam.

I spent June 1966 thru March 1967 with the 335th AHC at (Bien Hoa), flying with the Falcons.

I extended my tour by six months and, following extension leave, I was reassigned.

In April, I arrived at Cu Chi to be the aviation officer for the 1st Bde, 25th Inf Div.

We flew H 23 Ds. In late September, I was medevaced thru the 12th at Cu Chi to the 106th General Hospital in Japan, and further to Fort Benning.

I left Benning in February 1968 and had Flight A-11 at Fort Wolters until February 1969 when I received a F/W rating at Forts Stewart and Rucker.

I returned to Vietnam in June 1969 for five months with the HQ of the Avn Bde at Long Binh before taking command of the 335th AHC on 2 November 1969 (at Bear Cat). I departed the company 30 May 1970.

I was married on 14 June 1970 in Sausalito, CA and we drove to

Tallahassee, FL. where I taught ROTC for three years and added two sons to our family.

June 1973—May 1974 we spent at Fort Leavenworth, KS at command and general staff school.

June 1974—May 1977 was spent with the Southeast Region of the Recruiting Command where I was the operations officer and we lived in Stone Mountain, GA.

I was transferred to Forces Command Headquarters in June 1977 and worked in G-1 with the Recruiting and Retention Div. until I retired 1 August 1982. We had moved to Marietta, GA in 1979 and have remained here since.

Since retirement, with sons aged 10 and 11, I spent a great deal of time with scouting, schooling, fishing, sports (including SCUBA diving), summer travel, and working HONEY-DO lists.

I've also worked my way from being a pew-setter in church to be a worker, from trustee, to building program coordinator, to habitat house worker, mission trip worker, and handy-helper with a group when members need small jobs done.

I've also played a lot of golf and had very enjoyable visits with all who have made the COWBOY REUNIONS.

GOULD, THOMAS L.

I graduated R/W Flight School at Fort Rucker, January 1969.

Completed M-22 wire guided missile gunnery training at Fort Rucker, February 1969.

I arrived in Vietnam in March 1969 and was assigned to the 335th AHC at (Bear Cat), flying with the Ramrods.

April 1969 arrived Yuma Proving Grounds and was assigned to Army Aviation Test and Evaluation Command.

August 1971 Fixed Wing Q Course, Fort Stewart/Fort Rucker.

November 1971 Fort Huachuca, assigned to Fixed Wing VIP flight detachment at Libby Army Airfield.

April 1972 left the Army and went to work at Cochise Airlines, Tucson, AZ flying Cessna 402s and DHC-6 Twin Otters. I was married for five years.

June 1978 arrived in Pago Pago, American Samoa flying for South Pacific Island Airways flying DHC-6 Twin Otters.

November 1979 completed R/W External loads course at Aris Helicopters San Jose, CA in AS350 D.

January 1980 arrived Fiji Islands to fly AS350 Ds for Pacific Crown Aviation.

June 1982 Flying summer contract for Pacific Crown Aviation in Spokane, WA flying AS350 D in Montana. Received my A&P mechanics license.

November 1982 Flying for Healy Tibbits offshore construction in Santa Barbara, CA, flying Hughes 500, and AS350 D.

June 1983 working for Maritime Helicopters in Dutch Harbor, AK, flying AS350 D.

September 1984 returned to Spokane, WA Pacific Crown Aviation DBA Eagle Helicopters flying an air ambulance for Sacred Heart Medical Center. Completed EMT training. Married a flight nurse from the hospital in May 1988, who was killed in a helicopter crash in August 1989.

June 1990 returned to Fiji Islands flying AS350 D.

November 1992 quit flying and started farming in Northport, WA with my wife's father.

November 1995 moved to Destin, FL and built a home on the beach to retire. May 2001 married for the third time. Divorced November 2001.

November 2001 sold the beach house, and moved to Dothan, AL to play golf and enjoy my back porch.

GREENE, STEPHEN HOOD

After Vietnam, I returned to Fort Hood, Texas in the spring of 1968. I was initially assigned to the 2nd Armored Division—both the 1st and 2nd Armored Divisions and III Corps were based at Fort Hood—and after interviewing with Colonel Reed, who oversaw Aviation operations, I was assigned to the III Corps flight section. This was advantageous to me as III Corps had its full complement of aircraft, both fixed and rotary wing. Both the Armored Divisions had a single Huey each. III Corps had 3 Hueys, 2 OH-23s, 2 OH-6A, and various fixed wing aircraft. III Corps had only about a dozen pilots while each armored division had hundreds assigned to them. Colonel Reed noted my OH-13 time while flying in the Casper Platoon. So that the Armored Division pilots could stay current as pilots, he devised a plan to acquire numerous OH-13 E, G, H helicopters for them to fly. During that period there were very few Hueys flying in CONUS, outside of Fort Rucker and AHCs forming up to be sent to Vietnam.

I was sent TDY to Fort Wolters and went through MOI there in the OH-13. Upon completion, I was made the SIP for Fort Hood in the OH-13. My job was to develop a training syllabus, ground school, exams, and then check out many 1st and 2nd Armored Division pilots as IPs. They, in turn, would check out the pilots in their division. To do this, many dozens of Korean War vintage OH-13s arrived at Fort Hood. The oldest OH-13s was the E model with wooden blades. The H-13s was a sweet autorotating aircraft with the E model the best.

The pilots were all recently returned from Vietnam, so acquiring the basic skills in the aircraft came quickly. We worked on full-touch-down autos and other emergency procedures for the bulk of the training I conducted. I am sure I did thousands of autos during my tenure at Fort Hood.

Aside from the training, I also handled the job of pilot in the flight section. We had three Hueys, one which was a Presidential flight section Huey decked out in bright green, white and black, with red plush seats and wood paneling on the inside. It was there because President Johnson had his ranch nearby. We never flew anything resembling the president. We did support operations at Fort Sam Houston, Fort Bliss, and Fort Huachuca with all three Hueys. After Martin Luther King was assassinated and during the Democratic Convention in Chicago in 1968, we flew our Hueys to Chicago to assist troops sent there.

I was sent to Fort Rucker twice TDY. I first went through HIFC, the helicopter instrument flying course to receive my standard instrument card and then returned to go through the instrument examiners' course, to become an instructor and examiner in instruments. We flew precious little instrument time in the Hueys, since I was the only one in our flight section with a standard card. I flew lots of time in the OH-6s and even flew a lot of co-pilot time in our fixed-wings. I flew in our U-21s, Beavers, T-41s and even a C-45J.

While at Fort Hood, I acquired my commercial and instrument ticket in both fixed and rotary wing. I was offered a direct commission in the Field Artillery as a First Lieutenant but declined when I found out that I would be sent to Fort Rucker TDY to go thought AH-1G, IP gunnery school on my way back to Vietnam. Although in hindsight it might have been a good idea to make a career out of the Army, I never looked back and do not regret for one moment leaving the Army for good in January 1970.

After leaving the service, I returned to my home state of California, living in San Francisco. I enrolled in college as a freshman at the University of California, Berkeley, where I graduated with a Liberal Arts degree in 1974. I tried to use my considerable helicopter flying experience, over 2800 hours in the military, to find a job. There were no flying jobs at that time in the Bay area. At least none that I could do while also going to college.

I kept my flying skills by renting Cessna, single-engine planes mostly, to fly around co-eds that I wanted to impress. Within days of graduating college, I went to work for the FAA as a developmental Air Traffic Controller at Oakland Center. I stayed with them for a year, realizing very early that I did not want to be that close to flying while sealed in a windowless building looking at radar screens. Without a job in hand, I resigned.

Finding a flying job was only slightly easier than it had been four years earlier after leaving the Army. I interviewed with all the helicopter operators in the Bay area, but none offered me a position. I still had G.I. Bill left and enrolled at the Sierra Aviation Academy at the Oakland Airport. I used the remaining G.I. Bill money to pay for an external loads course they offered in JetRangers. Sierra did not have their own helicopters, so they contracted with an FBO at OAK to do that training.

After completing the training, I finally got some work with that FBO called Astrocopters—no pay and very little flying time, but it was a start. I washed helicopters, worked on the ground crew for lift jobs, and whatever else they wanted to keep in, around, and flying helicopters. I was given flying jobs no one else wanted to do and since they were not paying me, it was a bargain for them. I held other jobs to pay the bills.

While in college, I worked as a short-order cook and as a lifeguard at a nude beach near Devils Slide, just south of San Francisco. After the FAA, and while "working" for Astrocopters, I worked at night as a medical insurance claims adjuster, all to just pay the bills. Finally, I got a break: Astrocopters offered me a job as a Department of Forestry contract pilot on a Helitack crew to fight forest fires. All I had to do was complete the Bell factory ground and flight course, near

Dallas, in a JetRanger, all at my own expense. I spent all my savings and then some, but did the courses.

For the next five months after that, I worked twelve days on and two days off in the Mendocino National Forest. It was a fantastic job and used many of the skills I had from Vietnam. After the fire-season, I flew 'ash and trash' for Astrocopters, except now they paid me. I flew that unit the next fire season, where I worked at least the twelve and two schedule.

In the next off season, I got another flying job, this time with Commodore Helicopters out of Mill Valley California. They did many things, but their primary business at the time was flying tourists from their helipad at Fisherman's Wharf in San Francisco. I was only offered a part time position, but often flew more than full time. Most of the flights were four-minute jaunts from the Wharf out and around Alcatraz Island. I would do ten or more of these flights every hour, often for eight hours or more a day. I could do a take-off, landing, a flight over the bay around Alcatraz and eat a cheeseburger in four minutes. Needless to say, I got pretty good at take-off and landing.

When it moved back to fire season, I was a relief pilot on a fire contract. I would work three days flying tourists, fly the third night up to Redding, California, spend the night there. Drive early the next morning to Bieber, California to fly on contract for four days. The fourth night I would either fly back to the Bay area that night or early the next morning to resume flying tourists. I did that all summer and fall.

At the end of the fall, I was hired on full time at Commodore and shortly after offered the Director of Operations position. Along with DO, I flew the morning and evening traffic report for the leading radio station in San Francisco, as well as numerous other kinds of flying. I flew very little in our tourist division. I did fly for TV news, movies, commercials, and general charter for the next five years.

In 1983, I got wind that Stanford University would be starting up an EMS program, flying twin-engine helicopters. I put my name on the list to be one of the two pilots chosen. In April 1984, I was hired by Rocky Mountain Helicopters to fly on the Stanford contract. It was a 24/7/365 contract with only two pilots for the first eighteen months. I worked up to 72 hour shifts during that period, with either 48 or 72 hours off between shifts. It was a grueling schedule. They went to three pilots for about another year and then finally to an industry mandated four pilots per contract.

After my first years at Stanford, I was promoted to Chief Pilot. I worked at Stanford until June of 1998, when the company, now Air Methods, did not bid the contract and my job ended. Air Methods offered me a line job in Carlsbad, California, near San Diego, flying BK-117s and Bell 222s, still in an EMS role. I stayed there four years, commuting from my home near San Francisco. Four days on, four days off, alternating between 12-hour day shifts and 12-hour night shifts.

In 2002, I was offered a job flying 222s with CALSTAR's EMS program in Ukiah, California, about 100 miles from my home. That was a lot better than the 500 miles in Carlsbad. I took the job there and became their IFR training Captain and line pilot. Eventually, I became a check-airman for CALSTAR while continuing as a line pilot.

October 23, 2012, was my last flight as a PIC. I flew IFR up though a low cloud deck, transitioned to Night Vision Goggles for a scene landing, at night, in the mountains. I departed with NVGs, requested and received an IFR clearance from my old employer, Oakland Center. I shot an ILS approach to near minimums at Sant Rosa airport to deliver the patient to the hospital. I flew IFR back to Ukiah, shooting a precision GPS approach and stepped out of a helicopter for the last time. Nearly 47 years of flying, no accidents, no one hurt, and no violations.

After retiring I do whatever I want. I am life members of the DAV,

VHPA, 173rd Airborne Association, and the University of California Alumni association, and am a member and past President of the Casper Aviation Platoon Organization. I live with my wife, Carmen, of over 30 years, one cat and two love birds. I have one son, Skyler, who is now 27 and is a remarkable young man.

Carmen and I travel a few times a year. We usually attend the Casper reunion in June and a trip or two to the east coast to visit friends and family, as well as trips to Europe, Australia, and other spots.

In March 2017, I had the honor to return to Hill 875 as part of an investigative team from the Defense POW/MIA Accounting Agency. The 173rd still has three MIA from that battle that they hoped to obtain additional information about their whereabouts. Along with a team of active duty Soldiers, Sailors, and Airmen, I traveled with Jim McLauglin from the 335th AHC and Wambi Cook, a fellow soldier from the 173rd Airborne Brigade. We were told that we were the first Americans on that hill since the war ended and would likely be the last veterans from that battle to return.

Hill 875 and other areas are highly restricted with access to anyone extremely limited. Even the filmmakers from the recent PBS series, "The Vietnam War" were denied access to our old AO. This was an investigative mission, not a recovery mission. Our hope is that we provided additional clues to solve the mystery of their disappearance. Here is a link to the 19-minute film I made on the trip to Vietnam.

https://www.greenaire.com/hill-875-battle-for-dak-to.html

Please check it out and leave a comment. My home is in a canyon, only ten minutes from San Francisco, but isolated so that my neighbors are not visible from our house. We feed the raccoons and deer who congregate on our land. We see foxes, a bobcat or two, and the occasional mountain lion. Life is good. I have much to be thankful for and am still healthy enough to enjoy each day.

HENDLEY, GERALD W.

I joined the Army in 1966 as an enlisted Infantry soldier and went through Infantry OCS. I am from South Georgia and flight school was my fourth choice after airborne, ranger, and special forces. After the eye exam and fast OP test, the infantry was now aviation.

I joined the Cowboys in Phu Heip in 1968 and was a Ramrod for two months until Gale Smith gave me my chance in guns. Life changed. I saw what a gunship can do every day and I fell in love. I left in June 1970, leaving the best unit I have ever known.

Back to Mother Rucker as a flight instructor in Tactics.

I branch transferred to Military Police in 1972 and stayed on the ground ever since. Officer advanced course in 1972, instructor at the MP school at Fort Gordon, bootstrap at University of Southern Mississippi, and then to Germany for three years. Served as battalion S2/S3, Provost Marshal of Mannheim, and company commander of the 59th MP company in Pirmasens.

Along the way I got married to an Ozark, AL girl and have two daughters with her.

After Germany, I went to grad school at Auburn (War Eagle), Air Force Command and Staff College, and then to the Pentagon for three years. I was in the Army Secretariat, public affairs, as the press spokesman for budget, R&D, and weapons. I truly despise reporters of every shape and size. All are pissants.

Got divorced there. Went to 2nd Infantry Division in Korea as the Division Provost Marshal; loved every minute of the job. By the way, I was promoted to Major out of Germany and Lt Colonel in Korea.

Back to Atlanta as Chief of Plans and Ops for the FORSCOM Provost Marshal for two years, and back to Germany as Chief of OPS and plans for Region Commander, CID Europe and finally as Deputy Commander, CID Europe.

Remarried, had another daughter and retired.

Entering civilian life after over twenty-two years in the Army was an adjustment, but I went to work with Wells Fargo Armored Cars and it was the next closest thing to the Army. I was a branch manager in West Palm Beach, Miami, and Orlando.

Divorced again in Orlando and changed jobs. For twenty years I have run finance offices in Atlanta and Fort Walton Beach, FL until I retired two years ago. Now my daily schedule includes drinking wine on Tom Gould's back porch in Dothan, AL and looking at the golf course we play often.

I remarried ten years ago, but my wife lives in central Alaska and has for over six years. We get along well. All in all, life is good.

HENTHORN, TIMOTHY G.

Tim Henthorn is an experienced flight veteran with over 30,000 flight hours logged in various aircraft. He is known for being detail oriented and has a history for being thorough and efficient, which is supported by his accident-free record.

PROFESSIONAL EXPERIENCE

VIH Cougar Helicopters / Construction Helicopters | Boise, ID
Assistant Chief Pilot, 2010–2014

- Command Pilot / Check Airman Bell 212/205
- USFS Fire Contracts, Ogden, UT and Salmon, ID

San Joaquin Helicopters | Delano, CA | Line Pilot, 2009-2009

- Command Pilot Bell 212
- USFS Fire Contracts, Prescott, AZ
- USFS Fire Contracts, Chuchupate, CA

Swanson Group Aviation | Grants Pass, OR | Line Pilot, 2007-2009

- Command Pilot SA315B
- Magnetometer Surveys, Venezuela
- Fire Contracts and Seismic Work, Texas / Arkansas

EARLY CAREER HIGHLIGHTS

Copper Mountain Helicopters | Orem, UT | Owner, 1984-1988
Air America | Washington, DC | Helicopter Pilot, 1972-1973
US Army | USA | Helicopter Pilot / Flight Instructor, 1967-1970

HICKEY, JOHN JOSEPH

After Vietnam and leaving the Army in 1968, I went through several different jobs. In 1970, I was accepted into the Tucson Police Academy. After graduating from the Academy, I was a uniformed police officer for a year. After that I was transferred to the Tactical Squad which was a plainclothes assignment. We were assigned to work major crimes like homicide, rape, armed robbery, kidnapping, etc.

After four years of that, I applied for and was transferred to the Narcotics Unit. I worked deep undercover for four years, portraying drug users, drug dealers, and some organized crime. After leaving undercover work, I went back to uniform patrol, and did some time flying as an "observer" on the police helicopters.

While on the Tucson Police Department I enrolled at the University of Arizona and majored in Law Enforcement Management.

After 10 years on the police department, I left to open my own business, Aztec Orthodontic Laboratory, with my wife. After 46 years we are selling the business in January 2018 and retiring to our beach house in Puerto Penasco, Mexico.

In 1971 I married my beautiful wife, Vicki. We are still married today and have two daughters and four grandchildren.

I still think about my time in the Army, and Vietnam in particular. Good times and bad times. I guess it never goes away.

HICKS, RALPH DOUGLAS

After completing the tour with the Cowboys in Vietnam in early July 1967, I returned to my home state of Tennessee and married Arlene Coulter. We then moved to Fort Rucker, Alabama where I expected to complete my remaining military obligation as a helicopter instrument flight instructor.

In 1969, I decided to accept the Army's offer of a direct commission from W2 to First Lieutenant in Air Defense Artillery. I attended the fixed wing qualification course and returned to Vietnam and flew fixed wing aircraft.

Following the second Vietnam tour, I was assigned to the Army degree completion program and graduated from the University of Tampa with a BS Degree in Business Administration. After that assignment, I was committed to complete at least twenty years in the Army. Subsequent Army assignments included Ramstein AFB, Germany, Fort Bliss, Penn State ROTC, and Fort Rucker.

I retired from the Army in 1985 after twenty years of service. The Army years were challenging for Arlene and me, but we wouldn't change anything. It was a wonderful experience and we both took advantage of educational opportunities. She completed a Master of Science (MS) in Science Education and I completed an undergraduate degree and a Master's in Business Administration (MBA). We were also blessed with two sons Greg (1970) and Jeff (1973).

Immediately after leaving the Army, I began working for FlightSafety International in Daleville, Alabama. As an original member of the FlightSafety team in Daleville, I was fortunate to participate in the ever-expanding mission of providing fixed wing training to all the US military services and other government agencies. We also provided rotary wing flight simulator services to support the Fort Rucker helicopter training mission. My tenure with FlightSafety included the positions of simulator/academic instructor, Assistant Center Manager, Contract Manager, and Center Manager.

I retired from FlightSafety in 2011 after over 26 years with the company. Clearly, my early experiences with Army aviation had a profound influence on the decisions I made regarding my future working career. With a combined 46 years of working with Army aviation, I had a very satisfying career and enjoyed the continued military comradery with America's finest soldiers.

After retiring, Arlene and I have traveled extensively via cruise ships and European land tours. We continue to enjoy good health and have visited over 25 countries in the last six years. We strive to remain active as we approach our mid-seventies, so we can continue to enjoy hiking trips, antique car events, and numerous land excursions from cruise ships. We also enjoy spending time with our sons, their wives, and six grandchildren in Birmingham, Alabama and at Lake Martin, Alabama. (Roll Tide…War Eagle)

HOOKER, ANDREW E

After Vietnam, I was discharged September 15, 1970 from Fort Lewis, WA and returned to my parents' home outside of Boston, MA. Worked as a backstage security coordinator for a concert promoter based in Atlanta and spent the balance of the year working various rock shows and festivals from Orlando to Baton Rouge, LA. Had my first encounter with the VA, dealing with several health issues incurred while in the service.

1971 enrolled at Graham Junior College, Boston, majoring in Communications Media/TV and Film Production. Did project work for Music Unlimited and in 1972 became the Director of the Graham Student Center, coordinating student activities and entertainment events. During the summer and school breaks worked various shows around the Boston area and became an advocate for Vietnam veteran concerns.

1973 graduated with an AAS degree and transferred to the University of Massachusetts, Amherst. Became the UMass concert committee production manager, which presented the opportunity to further my experience coordinating production of large shows. Spent that year doing shows, and that summer left UMass and began working full time for Music Unlimited as road and production manager for several Boston rock bands. Continued through 1980 when my health situation sidelined my work for a couple of years.

1982 moved to Florida and enrolled at Eckerd College, St Petersburg to complete my Theater Arts Degree. Became involved with working shows again, at Eckerd College and around the Tampa market.

1985 joined Events International, Sarasota, FL as Executive Booking Agent. Later that year became the tour manager for the Country & Western and Bluegrass shows that were produced by the company. After several years on the road, I accepted the promotion to VP of National Tour Booking and Producer of the innovative Children's Travelling Museum, which crisscrossed the country with its positive and relevant messages for kids including the popular "Kids vs. Drugs" program. 1987 daughter, Ashley was born and 1992 son, Kamron arrived.

1995 sidelined again with health issues and spent the balance of the year recuperating. 1996 began working for Star Productions tour managing shows, including NFL Celebrity Hoops and the Harlem Rockets Comedy Basketball Show.

1997 was named Director of Marketing for Bristol Marketing Associates which was the marketing arm of Star Productions and the Motorsports Driver of the Year Award. This was a significant move because it was really the first time I wasn't heading off out of town on some road adventure. It was also during this time that both my children came to live with me. It was also the year that I reconnected with the 335th at the Las Vegas reunion. Amazing!

2002 seriously sidelined again with a heart attack, along with the other health issues that have increasingly gotten worse, began to seriously consider that invincibility may have run its course and started a route to retirement. That I did in 2005.

Since that time, I have stayed involved with volunteer activities involving veterans, active duty military, and their families. Have significantly curtailed these activities over these past few years; however, have continued to serve as the Vice Chair of the Sarasota National Cemetery Advisory Committee and 335th reunions. Relaxing while hanging out in my bunker in Ellenton, FL. My wonderful wife Cindy, has been doing all the heavy lifting.

HOPKINS, LANCE

Upon returning from Vietnam, I followed my older brother's example of hitchhiking home to my dad's gas station in Hermosa Beach, California, Craig was in country two years before me. Upon my arrival at Los Angeles Airport in dress greens (full uniform) and collecting my duffel bag, I proceeded outside to stick out my thumb and catch a ride to Hermosa. To my surprise, people drove by flipping me the finger and screaming all sorts of insults that should not be printed. When I left it was peace and love. In 1969 and my return was filled with hate and an atmosphere I had never known, not even in combat. Then all you try to do is blend in and rely on other veterans and friends for support.

I went back to my trade as a carpenter working for my uncle Paul as a rough framer. I completely enjoyed working outdoors and always moving to a new location not too far from home. I would stay in this occupation for my entire career, later forming my own framing company with a business partner Richard Mullin that lasted 32 years. I still travel across the Country helping people with projects and working with the tools of the trade (mostly family).

I have been married and divorced twice and have five children, eight grandchildren, and my present wife Linda has two children, one grandchild, and one on the way. The highlight of my life is my wife Linda, raising my children, and now spending time with the grandkids, what a true blessing!

My military experience certainly has had a huge influence on my life and on my world view of things. (I will not go political here).

The work ethics, discipline, and working as a unit in the Army instilled in me things that cannot be found in everyday life or perhaps taught outside of this environment.

I belong to four veteran organizations, the 119th AHC which meets annually, the 335th AHC which meets every two years, the Vietnam Crew Members Association, which meets annually, and American Huey 369 which has around fifteen events annually and is currently flying a helicopter I was crewchief on back in Vietnam in 1970, gunship 049.

HUIZINGA, DARREL BRUCE

After Vietnam I was reassigned to a Medevac Helicopter Company in Germany, just outside Munich, where I was a crewchief on a UH-1H Medevac helicopter from April 1969 until my discharge in August 1970. This was a very small unit with only two helicopters, eighteen enlisted, three commissioned officers, and six Warrant Officers. We were on an old World War II German Airfield. This turned out to be a really good way to spend my last sixteen months in the Army.

After leaving the service, I returned to Riverside, California, where in January of 1971 joined the San Bernardino Police Department where I spent the next six years. I earned a Criminal Justice degree from Cal State, San Bernardino, while working graveyard on the police department and going to college during the day. I eventually made detective in 1975 and stayed on the department until August of 1976 at which time I got married and left the department.

In 1976, I went to work as an Insurance Agent in Redlands, California, eventually becoming the owner of the agency in 1981. During this time, 1979, we had a son. The agency was very successful and in 1989 I sold it and became a McDonald's franchisee in Northern California. I eventually owned three McDonald franchises. In 2001, I got divorced, sold my franchises, and thought I would give retirement a shot. As it turned out, I remarried in 2003 and my retirement only lasted for three years at which time I opened an Insurance Agency in Marin, California, which I still own and run today.

I owe much of my success to the military. The military taught me discipline, teamwork, and the willingness to survive under adverse circumstances, which I have applied during my entire working career.

My two passions are boating and golf. I am a captain and have sailed in many parts of the world. I am cutting back on the sailing these days as it requires a lot more time than I can spend working at it. Golf, on the other hand, I seem to be able to fit in two or three times a week, plus my wife plays, which gives us playing time together. We do however own a ski boat. I just restored a 1981 Master Craft Stars and Stripes and I really enjoy taking it out on the lake. The only problem with a boat is what it stands for, **B**reak **O**ut **A**nother **T**housand.

To date I really have no plans to retire again. Owning a business gives me the flexibility to take a lot of time off. I have a staff of five which can run the business while I take off.

Just a side note, while in Palm Desert, California, a couple of years ago, I went to the Palm Springs Air Museum where they have a UH-1B gunship on display with all the Cowboy and Falcon markings on it. That brought back a lot of old memories.

HUNNICUTT JR., JACK J.

After Vietnam I was thinking life would be perfect in every way; instead I was in for a few surprises. I processed out of the Army in Oakland, CA after leaving Vietnam in October 1969 at the seasoned age of 19. Still just a kid in many ways, I was now knowledgeable of many things the average American could not envision. So, it was time for me to put the military behind and get on with my life. My memory of Vietnam began to fade, I didn't want to remember and sometimes those months in country didn't even seem real.

Knowing that your employer would hold your job open while in the service, I tried returning to my old job working at a grocery store. Within the first week, I realized earning a couple bucks an hour stacking shelves was a complete dead end. After paying for the gas and auto insurance, I hardly had lunch money left. My friend convinced me to apply for a patrolman position with the Atlanta Police Department and, to my complete surprise, I passed all the testing and was hired as a Police Officer at the age of 20.

Although I wasn't yet old enough to purchase alcohol, age was of little significance working the general patrol, evening watch. This was in 1970 during broad social and racial change in Atlanta. This wasn't a great environment for a five-foot seven inch, one hundred thirty-pound young man who just spent 19 months in daily contact with a M-60 machinegun. I had my butt whipped more than once, but never lost a perpetrator. Little did I know that my initiative and what some called a "hot dog" attitude, actually earned me the "right" to work the toughest wild west-like parts of the city. What a reward?

I survived five years in law enforcement, often thinking it was more dangerous than Vietnam, while clearly understanding there was no way I could stay alive at this job until retirement age. I moved to Florida and went into land surveying in what is now known as Palm Coast. The work was conducted in extremely remote areas, miles from the nearest road, and usually in waist deep water. Snakes, alligators, and mosquitos became my work mates, not to mention the ticks and leaches I took home underneath my clothing. It didn't take long for me to realize the swamps weren't much better than the streets of Atlanta or the rice patties of Vietnam. I needed a change.

Thinking back to when I was 17 and a high school dropout, I really wanted to join the Navy and follow along with my brother who had joined up five years earlier. To my great disappointment, the Navy recruiter realized I did not have a high school diploma and he directed me next door to the Army recruiter. And that is how I started with the Army. So here several years later, with Army and law enforcement experience, I decided to give the Navy another try. I had earned a GED while in the Army which opened the door for me to join the Navy.

I had just one priority in selecting a Navy career path, so I told the recruiter I must have a job with an active mission during peace and war, every month and every day of my enlistment. Reflecting, I really hated stateside Army life with the "make work" just to keep the troops busy. So, the Navy recruiter said this stuff called cryptology seems interesting, they work all the time across the world, but no one really knows what it is they do. Bingo! I'll take it.

The recruiter was correct! Naval Cryptology was demanding work, nights, days, weekends, and holidays, so the challenge I sought was delivered. Keeping tabs on our global adversaries was the primary mission and for me it meant more time in the Far East; to be exact I spent fourteen years in the Western Pacific aboard ships and shore duty. It was a great career working with an elite group of military professionals across all services. But as time moved on I found it a very natural transition to retire from active duty after 21 years of service and finally settle my family in Florida.

In the military I had gained experience with many different levels of automated systems and picked up a couple of college degrees along the way. I was able to leverage all of this into a civilian career in the software industry for the next 22 years. I was fortunate to have connected with a good company that offered me the opportunity for advanced education, leading to certifications from the Software Engineering Institute at Carnegie Mellon University. I really enjoyed this work but as technology frequently changes, I decided my life had been chocked full of change. And as Homer Simpson said, "When I learn something new, I have to push something old out of my head." I was ready to seek a simpler life and to keep busy under my own management, and at my own pace and schedule. Retirement has been great.

Military life was crucial in developing my ability to focus on goals, take on challenges, and adapt to whatever situation I confronted. I would not change one thing in all my experiences. The down times were always over shadowed by the great times and, as my father said, "Hard work has never killed anyone." The military will always hold a special place in my heart and the American Veteran will always have my untiring support.

JACKSON, CHARLES F.

After Vietnam, I was assigned to Cobra Hall, Hunter Army Airfield, Savannah, Georgia, as an Instructor Pilot teaching aerial gunnery. I left the service in May of 1969.

I returned home to Spokane, WA where I used the G.I. Bill to obtain a Master's Degree in Speech Pathology and Audiology from Eastern Washington University.

I spent the next few years working as an Audiologist with several ENT Physicians in Spokane, Colorado Springs, CO and Atlanta, GA. In 1980, the US Navy established an Audiology Program and I became the third Uniformed Audiologist to join.

My first duty station was Naval Hospital, Bremerton, WA as Staff Audiologist. From there I was stationed as Head of the Occupational and Preventive Medicine Department at Naval Hospital, Yokosuka, Japan. From there it was on to the Naval Medical Center, San Diego as Head of the Occupational Audiology Department. I retired from the Navy in 1995 as a Commander.

I continued working at NMC, San Diego until full retirement in 2006. Obviously, the military became a major portion of my life. Along the way, I married, and we have two children. A son who works in the financial business and a daughter who retired from the Coast Guard as a Chief Petty Officer. I have been married almost 45 years.

After retirement, we continue to live in San Diego. Since 2006, I am working 14 hours a week as a volunteer with the La Mesa, CA Police Department. I will continue to do this until it is no longer fun.

I am a life member of the 173rd Airborne Brigade Association, and a member of the US Coast Guard Auxiliary.

KELLY, JEFFERY R.

After Vietnam, when I first got home, I signed up for unemployment for a brief time, then went to work at a local factory and got together with musicians I used to play with before I was drafted. We played at local taverns and dance halls for a while. I played base and piano. For those who might remember, two pilots (I don't recall their names) and another maintenance guy we called "Frog" formed a band at Bear Cat.

My band at home broke up and I decided to go back to school. I attended Southern Illinois University under the G.I. Bill for about two years, then I decided to transfer to Colorado University in Boulder. I came home to Alpha, Illinois to visit that summer and met my wife (smartest girl I ever dated) and never went back.

I went to work at JI Case Company and eventually retired from there. I have always tuned pianos as a part time job, and I continue to do this. Now I am sometimes busier than I want to be, piano tuners are becoming scarce these days.

I have been married almost 45 years; I have two kids and three grandchildren. I have been a "lunch buddy" with Big Brothers and Big Sisters. My wife was Catholic, and after we were married I eventually joined the church (one of the best things I ever did).

As far as the effect Vietnam had on me, when I attended SIU, being a Vietnam vet was the last thing I wanted anybody to know. The atmosphere at school was totally antiwar, antimilitary, and antigovernment. It just wasn't cool at that time to be a vet, and that feeling stuck for a long time after I got home. To be honest, I wasn't happy about getting drafted, but my dad and uncle were both honored World War II vets and there was no way I could disgrace my family by avoiding my military obligation.

KENDRICK, LARRY (KEN) DALE
Submitted by Joretta Kendrick

Before Vietnam, Larry worked as a driver/mover in his family's moving business. After Vietnam, he went back to work at his family's moving/trucking business for a while. Once leaving the family business, he became an over the road coast to coast truck driver. He did this for 26 years. At one point during his trucking career, he became an owner/operator and had a dedicated run, hauling mail from Cincinnati, OH to New York City. He did that for a few years and then went back to driving coast to coast.

In 1985, he had health issues and left truck driving and became a diesel mechanic, he did this for four years. After that, he started working at a fireworks company and became a licensed pyrotechnician. This was his dream job, he loved blowing up things.

In 1994, he left his fireworks job to start working for the state of Ohio as a prison guard. Even though he did this, he never gave up his fireworks and he continued to do firework shows for many years in his spare time and as his health would allow. He worked for the prison system for five years but had to take a medical retirement in 1997 due to injuries he sustained at work.

In late 1969 or early 1970, Larry married Nancy and they had three children, Michael, Andrew, and Keri. They were married for about 14 or 15 years and then divorced. After his divorce, Larry was single for a bit but did remarry. We (Joretta) were married for 28 years, up until the time of his death, May 5, 2014. While we never had children together, he did raise my two children, Jami and Brandon, as his own.

After the war, Larry's family said he became much more patriotic and wanted everyone to know what a privilege it was to live in the USA. While Larry did not talk about Vietnam very much, the effects of the war forever changed him. He spent a large part of his life hiding the pain of Vietnam behind alcohol.

Things changed for him in 1992 when our son's teacher asked him if he would care to share some of his experiences from Vietnam with the class. She told him that she didn't feel like their history book had enough information in them for the children to get a good understanding of what that war was really like. He agreed. He took in his medals, pictures, and other miscellaneous items that he had and did a question and answer session with the kids. He told the kids that he would answer the questions honestly and to the best of his knowledge.

He continued to do this for many years and it was moved from the class room to the auditorium so that the whole fifth grade class could participate in the session. While Larry opened up a little more about his experiences and tried to focus on the good things, he never got over his dislike of Jane Fonda.

After Larry's retirement, he enjoyed travelling, spending time with the grandkids, family, and friends. He also became more active in his church. He participated in a weekly men's meeting and went on a yearly hunting trip, as the cook. He was not a big hunter, but he loved cooking. We also took a couple trips out west and visited some of his Army friends, which he really enjoyed.

Other hobbies he had were reading, watching movies and documentaries on other wars, and collecting coins and various weaponry. His health really started to deteriorate in 2011 and it became increasingly harder for him to do very many things. Larry was not a quitter and even though it was extremely painful and difficult for him to get out and go to the grandkids activities, family get-togethers, or just do whatever was needed of him, he would push through the difficulties and do it with a smile and little to no complaining. Larry loved God, his family and friends, and his country greatly. He has been missed deeply by all that knew him, May 23,1947 — May 5, 2014.

KINNAN, MORRIS MERVIN (MERV)

I left Vietnam on November 7, 1970 after an 18-month tour which included a six-month extension that granted me a year and a half drop off my active duty. So, I was virtually out of the Army as soon as I stepped back on US soil, because I was put on inactive reserve, which had no compulsory meetings or drill requirements.

As I boarded my "Freedom Bird" that day, I felt a great emptiness and sadness, as well as a deep feeling of abandonment of my comrades in arms. I even toyed with the thought of requesting that I be allowed to rejoin them before it was too late to do so. But just as rapidly as I was overcome with this emotional response to my leaving, the feeling of high expectations of seeing family and friends washed over me as I took my seat among the others that were leaving that day. I don't re-member if I even said a word the whole trip back to the "World." It really didn't matter to me because I had nothing to say to anyone any-how. I was melancholy and happy at the same time, but most of all, I could hardly process internally that I was going home alive without any visible injury. I guess you can say, I was in shock and disbelief.

After being processed out of active duty in Oakland, I headed for the SF airport for my flight to Seattle which was the nearest airport to my home town of Shelton in Washington State. I contemplated my future as I sat in a bar, awaiting my flight departure, and came to the realization I had not made any plans for my life beyond Vietnam. I also realized I no longer resembled the person I was before I left home to join the Army. I have become a man with narcissistic appetites that were totally opposite of the moral goals I had sought for myself before

Vietnam. Maybe after I go back home to familiar surroundings, I can redefine myself and become that person I once dreamed of becoming. So, I downed my drink and headed for my flight to Seattle.

Back to Square One.

After returning to my hometown, it became apparent I was not the only one that had changed. Some of the mom and pop stores had gone out of business and big chain stores started showing up. The feeling of personalized attention has not only departed from the stores but from the community in general. I don't even feel welcome on the streets anymore, especially if you divulge where you have just returned from. Old friends have moved away and those that haven't can be located in the bars every night, including me. I realized it was time to move on and take control of my life. I can't roll time back and become the person I use to be. It is time to drag this new baggage along and find a new goal to shoot for (no pun intended). But I sure do miss flying those B models and feeling I am in control of my destiny.

I got my commercial helicopter license and then my fixed wing licenses for single and multi-engine, as well as instrument rating. But there were just too many pilots out looking for jobs, so I decided to go to school. I stuck it out for a year and a half, but just couldn't stand feeling I was back in high school. I met my first wife in college and we started a family shortly after we met. So, I went to work to earn a living and soon was working in a coal-fired power plant as an operator. It was tough hours, but a steady job with good benefits. I continued to look for a corporate flying job and even worked on my days off from the power plant as a flight instructor at two different local airfields just to stay current.

Early in 1978, I started working for a family owned forest products company as their helicopter pilot. Everything was working out well until the recession of 1980 came along. Forest product companies usually get hit early in times of recession, so my corporate helicopter position ended around October 1979. I had some savings, plus I was still working as a flight instructor, so I decided to sit down with the FAA and write my own Part 135 operating manual and started Kinnan Aviation as a single pilot helicopter company. I leased a Hughes 500C from a local owner and hung out my shingle.

By March of 1980, I was flying ABC network out of LA around Mount St. Helens as it started to rumble and make threats of erupting. Very few changes were taking place on the mountain through the month of April, so the number of aircraft flying in support of the USGS and news media had diminished greatly. We were all getting tired from all the flying, so on Sunday morning, May 8, most of us decided to meet for breakfast at the Red Lion Inn on the Portland side of the Columbia River.

About 8 am, we felt the building shake and we all looked at each other as if to verify what we knew was happening. Mount St. Helens had just erupted and none of us were in the air. Within thirty seconds, we were all up and running for our cars to get to our respective aircraft that were at Pearson Airport on the Washington State side of the Columbia River. Well, the rest is history as we all shot footage of the devastation for the next couple of months. That has become so familiar to each of us as the news media replays this footage every May 8th anniversary of the eruption.

For me, that was a very lucrative time. But the helicopter owner saw an opportunity to make a killing and wanted to raise the hourly rate we were charging. When I refused to raise the rate, he withheld the maintenance on the helicopter and I had to park it when the 100-hour inspection came due. Because it was the start of the summer, all available helicopters were under contract in my area, so this effectively forced me to close my company doors and go back to work at the power plant.

As devastating as this was, I was not prepared for what happened within weeks after losing my helicopter business. My wife decided to pack up our two children and filed for divorce. While I wasn't totally unaware of her disgruntlement within our relationship, I figured I could make amends when the work situation settled down. But that wasn't going to happen because she had already found someone else.

As the divorce proceedings dragged on, I became more and more frustrated with my wife running around town with her new friend while the courts gave her free will to dump our children off with other people to care for them while all the time I wanted them home with me. I had often planned her and her boyfriend's demise, but I also knew full well that if I did kill them, I would be depriving our children of both their parents, one dead and the other in jail.

So, I decided to fight for the custody of our children who were eight and five at the time, even though I agonized over how I would be able to care for them and work at the same time. So, one day in the empty living room of our house, I fell on my face and asked God to come into my life and help me raise these children. I told Him I would stay single for their sakes and not seek a mate until they have left home. I promised to raise them up in a God-fearing environment where I will honor Him.

God answered my request and things started to happen. My old boss had taken a job across the state and called me up to offer me a new job, which I accepted. The child custody evaluations being conducted in the process of the divorce gave me full custody of the kids. I put the house up for sale and it sold within thirty days. The three of us were gone across state within 45 days of the divorce decree.

These events made me do a lot of introspection and self-examining. I wasn't sure that I ever loved my first wife, maybe I just liked the marital pleasures that were available. I suffered from what I called the 'Echoes of War.' The visions of the carnage I caused upon the human beings that were unfortunate enough to be within the range of my weapons. And I still harbored a hatred for a government that prevented us from winning a war that we could have won.

Over the next 16 years, God worked on my problems as my children grew up and moved away to college. My heart was softened and the 'Echoes of War' subsided and finally went away. I had a chance to move back to the west side of Washington State and work at the same power plant I was at before. My ex-wife had moved away so I didn't have to worry about dealing with her. But now I had the privilege to help with the end of life of my father and then my older brother, which I would not have had the opportunity without moving back.

The best thing that had happened to me since then was my meeting and marrying my present wife. She lived in Alaska when we met online, so after a year of our alternating between who traveled to visit the other, I finally popped the question. I moved her down to the lower 48 in 2006. In the meantime, a Canadian firm bought the power plant I was working at and we had an opportunity to go to Calgary to work at the home office. So, in 2008 we moved to Canada. After working there for three years, we moved to the Edmonton area to work at their coal fired plants in that area for my final three years before retirement in 2014.

In June of 2014, we moved to Alaska where we presently reside on five acres outside of the city of Wasilla. I also bought a place back in the woods with no roads and a summer cabin. In the winter, it takes a snowmobile to cross the frozen rivers and swamps to get there, or a float plane to land on the lake in the summer. I hope to spend much of my time out there doing some fishing and enjoying the peace and quiet. But then again, if the grandkids come out, there might not be so much peace and quiet.

What the Hell…at least they're not shooting at me out there.

My best wishes to all the Cowboys and their family members.

KRUTUL, ROBERT P.

When I left the Cowboys at the end of 1968, after one year of great flying and camaraderie with the best guys I ever had the pleasure to know, I extended and stayed another year and a half (three more tours!). I got into a VIP unit, the 201st Red Barons in Nha Trang. I went from flying every day to flying once every three days and supporting the general in II Corp with a Huey that had U21 captain seats and a red carpet. Are you kidding me? For the most part we had to fly a few of those colonels around, and in my last six months, things got so bad we were flying lieutenants out to fire support bases again!

I finally left Vietnam in April 1970 and decided the only thing I enjoyed was flying. So, I went to a flight school in Clearwater, Florida. I spent a year obtaining all my civilian and commercial helicopter and fixed wing licenses and a fixed wing flight instructors license. I then headed home to Chicago, on the south side where all the poor folk live.

I knew I wanted to make a career out of flying and decided that I also needed to finish college. I found the perfect university only 50 miles from home, called Lewis University, in Romeoville, Illinois. The university offered the ideal degree, a four-year bachelor's degree in aviation maintenance management which also included an A&P aircraft mechanics license.

I started in August 1972 and finally graduated in January 1975. Believe it or not, the school was located on a small non-tower uncontrolled airport and I was able to flight instruct and make a few bucks and slowly build fixed wing flight time. Getting into a company that had an aircraft and being a "professional" pilot took a long time. I spent the next few years just flight instructing in single engine Cessna's and a little in twin engine aircraft. In 1976, I was able to get into the Illinois Army National Guard and was flying Hueys again. Glory Alleluia! It felt so great to fly a Huey again and be current flying helicopters.

Finally, in 1981, I received my big break and was hired to fly a twin engine Agusta 109A helicopter out of Cleveland, Ohio, for Chessie Railroad as part of a flight management company. I was in Helicopter Heaven! I was finally in Corporate Aviation! It had two Allison turbine engines, retractable-wheeled landing gear, and an Auto-Pilot! YEP, an AUTO PILOT! Not Force Trim. Look Ma! No Hands! The flight management company had a big selection of multi-engine aircraft, a few turbo-prop planes, and a lot of jets.

They noticed that I had all my fixed wing ratings and multi-engine rating, so, besides flying the Agusta helicopter, they had me flying co-pilot on King Air 100s. Then, they had the courage to put me in the right seat of a jet! Holy Mackerel! I was building that rare JET TIME!

So, there I was in Pilot Heaven. Then Chessie Railroad decided to move to Baltimore. Then in 1977 they sold the Agusta and transferred me to Richmond, VA, where I ended up flying a Sikorsky S76B. Are you serious? Holy Cow, talk about a helicopter with power! Two Pratt & Whitney engines and retractable wheeled gear and of course an Auto-Pilot. But not just any auto-pilot….an auto-pilot with Auto Hover! Did you ever hear of such a thing? Oh, if we only had a thing like this in Vietnam. By the way, I also transferred to the Virginia National Guard to make sure I was getting my Huey time.

Well, as I found out, these great jobs don't last long. I was laid off in 1988 and went job hunting. But this time I had good experience to flaunt! Well, a few months later I fell into a bed of roses and got a job with the giant company E.I DuPont in Wilmington, Delaware. Oh my, I did it again. They had a Sikorsky S76B also. So, I was flying that, and they also had about 10 jets, so they put me into the right seat of a Hawker 800 jet. And, as luck would have it, the Delaware National Guard was on the same airport so naturally I transferred to the Delaware Army National Guard because I could not, in any way, give up flying that Huey! So, I did this for the next five years until the economy went sour at the end of 1993 and got laid off. Why do all good things have to end?

Well, by this time I was married and had a daughter, Allison. I did not want to be a long-distance dad, so I stayed near Baltimore and got a job selling cars! What else can a pilot do? But I sold Volkswagens, so it was ok because I got to love VW's and bought the car that I will never get rid of...a 1996 VW Jetta Trek! I still have it after 22 years.

Ok, I finally got back into flying a year later and said goodbye to selling cars (horrible job) and ended up flying a Hawker jet out of Harrisburg, Pennsylvania. But that wasn't the best part. They also had a three-engine jet, a Falcon 50, and a year later I was flying captain and flying over the Atlantic to Europe. My guardian angel took care of me again. Now, I can say I have flown to England, to Paris, and Rome, and Barcelona, Spain! Now, are you ready for this? I have even flown north of the Arctic Circle! Of course, it was August, but I still saw ice-bergs from 40,000 feet!

Well, that glorious job ended after three years because my wife had a bad knee injury that required moving to permanent warm weather because winter was killing her. So, in 1998, I moved to Phoenix, Arizona, got reunited with my brother and most of my family who migrated there years ago from Chicago. I did live in Palm Springs, California, for a few years flying that wonderful Hawker 800 jet again. But, again, a good job ended because the company was bought out and the airplane was sold. So sad.

Now I am living in the Phoenix area again in the most wonderful and best paradise ever...a secret paradise called Fountain Hills. It's next door to Scottsdale on the East side. We have our own man-made lake and the world's 3rd highest water fountain! Look it up! It goes up for 15 minutes every hour. Pure serenity.

I was laid off from my last job in April 2017 and just turned 73 and haven't found another job yet, so I might be retiring soon. Thing is, I love flying too much to give it up. I still pass my flight physical so I'm going to keep trying. In the meantime, I have finally got the time to finish building the Lionel Train layout that I have wanted to do for the past 50 years. Also, I am finishing up the construction of a radio-controlled model ship. Four feet long, called The N.S. Savannah, the first nuclear merchant marine ship. It has been in dry dock since 1963 because life got in the way. I hope to have it sailing in another four months.

Well, fellow Cowboys, that's my story. It was super nice to join the reunion in Las Vegas, I think it was 2000. I will try to make the next one if we have one in 2018. By the way, I retired from the guard in 1999 and stuck it out for 23 years with various National Guard units and am so happy I did. Nice to have a little retirement from all those Huey days.

I look forward to meeting all the Cowboys that are still in touch. It has been nice to be in touch with my old Vietnam roommate, Dave Fraser and John Shelstad. Hope to see everyone else real soon. Take care Cowboys!

KUHBLANK, RICHARD CARL

After leaving the Falcons in 1967, I was sent to the Engineer Officer's Advanced Course at Fort Belvoir (Davidson Army Airfield), Virginia. While there in 1968, I was pulled out of class one day while the Washington DC/Georgetown 1968 race riots were in full swing and cranked up a Huey B model gunship to fly HOT right down M street in Georgetown where the International Safeway grocery store was burning down. Fortunately, we didn't have to engage any targets!

From Belvoir, I was sent to Fixed Wing Transition Course at Fort Stewart, Georgia, and then on to the University of Southern California and the Army's Aviation Safety Course (where at lunch time I and several classmates ran around the USC track with O.J. Simpson). Then it was back to Vietnam for a second tour as the 1st Infantry Division Aviation Safety Officer. Neil Stribling (my old Falcon hooch-mate) was with me on this tour in the 1st Aviation Battalion (Phu Loi). While in Vietnam that tour, I received notification that the Army was sending me to graduate school at Purdue University in West Lafayette, Indiana.

I received a Master's in Geodetic Engineering at Purdue and went from there to become an instructor at the United States Military Academy, West Pont, New York (not a graduate, but taught military tactics, mapping, orienteering, and flew helicopters for the cadet sky diving team). Assignments after that included Germany (Engineer Topographic Battalion), Fort Belvoir, VA (Defense Mapping School), Air Command & Staff College (Alabama), Belgium (SHAPE—2 tours) which included project engineering for conventional and nucle-

121

ar facilities in countries from northern Norway to eastern Turkey and all western countries in between—special work in anti-blast and ballistic design. I also got some Huey time one day, courtesy of the Greek Air Force flying between the islands of Crete and Santorini, then back to Washington DC, the Defense Mapping Agency (Naval Observatory grounds) and retirement September 1, 1986 after 24 years of active service and advancing through the officer ranks to Colonel.

I then began my second career with the US Department of State (another 15 years) as a security engineer working for the Diplomatic Security Service and the Office of Foreign Buildings. This work included on-site assessments of security vulnerabilities at existing US Embassies and Consulates, assessing locations for new ones, and follow-up coordination with architectural designs to integrate important anti-blast and anti-ballistic features, hardline walls, perimeter barriers, safe areas, and electronic sensors and cameras and controls secured in protected Marine booths. Travels were extensive and included Japan, China, Mongolia, Vietnam, Cambodia, Myanmar, Indonesia, Australia, New Zealand, American Samoa, New Guinea, Colombia, Barbados, Chile, Guatemala, Kenya, Ethiopia, Egypt, Liberia, Zimbabwe, Madagascar, Mozambique, Germany, France, The Netherlands, England, Luxembourg, Sweden, Finland, Denmark, Greece, and Czechoslovakia (before splitting up).

I also visited Russia, but it was unscheduled and occurred when on the way from Beijing, China to Ulaanbaatar, Mongolia when a sandstorm over the Gobi Desert rose up to 36,000 feet high and forced my plane to land in Irkutsk, Russia where the Russian military held the plane and its occupant's captive for several hours while seeking special payment before releasing them to fly on to Mongolia. Later, I celebrated near Ulaanbaatar with local tribesman in their skin covered Yurt while drinking fermented mares milk (great stuff).

I was also a member of one of the first US teams to visit Hanoi after the Vietnam war. We suffered from cancelled meetings, broken elevators, and criticism, but still established a temporary US Embassy until a new one could be built. A major section of the "Hanoi Hilton" prison was being demolished by the Vietnamese government and I refused to visit Ho Chi Minh's mausoleum. In the south, we demolished the old US Embassy in Saigon (Marine sandbag emplacements still in position on the roof) and replaced it with a new US Consulate (complete with modern security features). The old 173rd Bien Hoa base camp was gone — only rice paddies and scrub jungle again!

Cambodia was also US friendly after the Vietnam war and as I walked through the tiny air terminal in Phnom Penh to begin upgrades on the US Liaison Office to bring it to Embassy status, I walked right by the same room where during my tour with the Falcons, I and one other G.I. had been erroneously booked on an R&R flight to Hong Kong that made an illegal stop in Phnom Penh to deliver communist NVA agents. We had been forced off the plane at gunpoint and forced to view a black and white propaganda 20-minute film before being allowed to get back aboard our flight to Hong Kong. US Air Force intelligence service later shut down this system of transporting agents after I reported the incident.

My experiences during my second career were rewarding, sometimes dangerous, exotic, and physically and mentally demanding. I worked with tremendous dedicated, professional people who spent their whole careers forwarding US goals as a world leader for peace and freedom. 97% of these folks were dedicated to their work and only about 3% just came to work. Things got a little unnerving when you knew you were being "tailed" or under electronic surveillance (China, Laos, Vietnam, Indonesia, Colombia, Madagascar, Ethiopia, Egypt, Liberia, Finland, old Czechoslovakia and East Berlin before the "wall" came down), but you needed to remain calm.

I retired from this second career in 2002, sold my home in Springfield, Virginia, and headed for the Delaware beach. Ocean View is small, historic, and a 10-minute bicycle ride to the beach. I now spend my time with family (everyone loves beachy things), church (Mariner's Bethel United Methodist Church where I still sing in the choir), golf, VFW (Post 7234—Life member), reading, traveling, and sleeping soundly. I wish the latter to all the Falcons and Cowboys who have taken the time to read this. In summary I am one contented dude!

Family information:

Richard Carl Kuhblank. Born: May 24, 1940 in Everett, Washington (north of Seattle). Raised in Mount Vernon, Washington. University of Washington (Seattle) graduate, ROTC commission, Sigma Chi Fraternity (Life member), and Purdue University (West Lafayette, Indiana) post graduate.

Lene Kuhblank. Wife. Born: (she's younger than I am) in Copenhagen, Denmark. Raised in Taarbaek, Denmark (beach/marina town just a few klicks north of Copenhagen). Interior decorator, mother, banker, historian, church coordinator (and love of my life).

Christopher Scott Kuhblank. Son. Born: May 13,1970 (while Dad was at Purdue). Marymount University in Washington DC and George Mason University in Virginia. Teacher, coach and assistant athletic director at Madison High School in Vienna, Virginia. Married Gena Ann Clare in 2000 in Bermuda. Children: Ryann (girl) Kuhblank, September 13, 2001 in Arlington Hospital (while 9/11 Pentagon victims were being treated on another floor). Bo Kuhblank June 26, 2003 in Arlington Hospital.

Rich met Lene on July 4, 1963 near Heidelberg, Germany. She was in a summer language course at the University of Heidelberg. They were married in Taarbaek, Denmark on October 25, 1963 in a dual language ceremony followed by a carriage ride to the Danish King's Hunting Castle for an elaborate formal reception (right out of Disney). Many tremendous years together (you do the math)!

The name "Kuhblank" is German. There is a small town in Germany named Kuhblank. It is in old East Germany on the East bank of the Elbe River near Wittenberg on the main railway line from Berlin to Hamburg. Lene (fluent in German) and Rich visited Kuhblank in 1993 after German reunification. This was the first contact with two remaining Kuhblank families since 1871. I met another Richard Kuhblank (in his 80s — now deceased). The small hamlet (no more than 130 people) with a circular main street was still a farmland and I was asked to buy an abandoned farm (the young Kuhblanks had gone to better their lives in the Western part of Germany), but didn't bite.

Exit: When I get liftoff for that final eternal flight, you can find me at the VA Cemetery, 26669 Patriots Way, Millsboro, Delaware 19966. Thanks for being a part of my exciting life. Peace be with all of you, always! RICH — Falcon 87

LA BRIE SR., ROBERT EDWARD

I stayed in the Army until I completed 22 years and then retired.

I left Bear Cat on 1 January 1970, returned to Portland Maine, got rid of my cheating wife and was assigned to the 2nd Aviation Detachment at West Point, NY, but it was actually at Stuart AFB in Newburgh New York.

I was assigned as a maintenance Supervisor, tech inspector, and First Sgt. so I wore many hats. After being assigned there, I married my wife Ellie. I left there in November and was sent to Camp Casey, Korea as a Battalion Maintenance NCO. From there I went to Fort Devens, MA as a 1st Sgt of the D Troop 10th Special Forces Bn. Then was assigned as the NCOIC of the Army Airfield at Fort Devens. I left there in 1975 and was assigned as the First Sgt of Company D of the 725th Maintenance Battalion. My next assignment was the 68th Dustoff Det. as First Sgt. Left there and came to Bangor, Maine as the Senior Enlisted Aviation Adviser for the State of Maine, and retired June 30, 1982 with 22 years of service.

Got divorced in 1970 and remarried in June of 1970, I had a six-year-old son from my first marriage and my wife had a six-year-old daughter from her first marriage whom I adopted. Together we had a girl and a boy, so we have yours mine and ours.

I have really missed the service since I retired. The military has always been a very big part of my life. Most of my friends are military, some are lifers as they call me, and some are my brothers from my two tours in Vietnam. I also had three cousins that retired from the military and one that was killed in Vietnam.

After retirement from the Army, I worked as a night supervisor for the United States Postal Service for eleven years. After that, I worked for Airborne Express for six years, then Uniship Courier Services for five years. I retired again at age sixty-six. I thought I was done working, but after sitting around for about three months, I went to work for a local Kia dealer and its been fifteen years now and still going.

I belong to the Disabled Veterans of America, the Purple Heart Association, The American Legion, the Veterans of Foreign Wars, and my favorite, Patriot Guard Riders of America.

LAGRELIUS, ELMER LEON (LEE)

Served four tours in Vietnam—60 & 61 with MACV (fixed wing pilot), 64-65 (was transitioned to helicopter) served with 1st Infantry Div., 1967, 1968, 1969 (335th), 1969-1971 I taught at Fort Rucker, AL. 1972 in Vietnam with MACV.

After Vietnam, I attended Slippery Rock University in PA for a Master's Degree in Psychology.

I owned an HVAC Company for several years. I was contracted by American Air Evacuation (DoD) and subsequently joined the Pennsylvania Air National Guard.

I was hired as a Social Worker at the Butler VAMC, Butler, PA from which I retired after 18 years in that position.

I was married three times. The first time I was married for seven years (and had twin sons), the second time for a year and a half (no children), and the third time and still going strong for 30 years with two great stepchildren.

At the time, regarding the world situation, I thought the Vietnam War was necessary and the US intervention was warranted. I later became disillusioned when we had numerous opportunities to win it, however we were not permitted to do so when it became politically unpopular to win it.

I was very angry about how we Veterans were treated upon returning home and furious about the number of men who fled to Canada then could return to the US with no punishment nor repercussions whatsoever. At the very least they should have been required to provide three years of public service. At the same time, those of us that did serve were treated like crap, shunned, and looked down upon.

I do not believe in a draft as there are other ways to get men and women to enlist in the military. This is now being proven by the number of people joining the military for the many opportunities they are being afforded.

I spent 23 years total in the Active Army and PA Air National Guard. My Health was generally good until age 65. At that time (2005) my health issues started to decline in earnest. I had AVR replacement, a pacemaker, a stent, colon, thyroid, and prostate cancers. The VA determined these were due to Agent Orange exposure. I was exposed daily (and plenty) as a gunship pilot flying gun cover for Ranch Hand (the main Agent Orange sprayers). I believe I got a larger dose due to this assignment. Those of us that were gifted with that duty were granted SC disability after discharge. I received compensation for several years. My heart issue (also determined to be Agent Orange related) increased SC to 50%. After the stent it went to 60% then several (numerous) other medical conditions, including PTSD, went to 70%. Two years later when the cancers struck, all of which prevented me from working, I was awarded 100% SC Permanent and total. I would rather have good health, but since I don't have much choice in the medical matters that befell me, at least the compensation insures my wife and me a steady income and comfort for years to come.

All in all, the military afforded me a good career, but I can honestly say now that I could have done without Vietnam. While I was trained as a pilot (fixed wing and helicopter), I had a short stint as a Commercial Pilot, had a small business as a Charter Company, and an airplane parts business.

I was a member of several recreational flying clubs (until my heart surgery grounded me). I continue to this day to maintain my CFI certification as an examiner. I am a life member of the VFW and DAV and a member of our Community Veterans Club.

I have attended several 335th reunions as well as the recent dedication service at Fort Rucker last year. During the reunions I have heard many derogatory remarks and complaints about the VA system. As for myself, the VA employed me for 18 years. I have received exceptional medical treatment from the VA and do so to this day. I was identified as an individual with traumatic injuries sustained from Vietnam combat as well as Agent Orange exposure and was duly SC for those life altering conditions.

After retiring from the VA in 1996, I enjoyed retirement doing things I liked to do but didn't have time to do when I was working. Due to my traumatic arthritis from my injuries, the Pennsylvania winters became too much to endure. I became a Florida "Snowbird" for several years in the harshest part of Northern winters. While in Florida I was able to continue doing some helicopter flying and was employed by a Fixed Based Operator in Clearwater. I had some interesting and enjoyable temporary jobs, including flying in the Anheuser Busch Copter and working for a heavy lift company setting beams and cranes for a Clearwater Harbor bridge and overpass project.

My wife had a rewarding career with the VA and was still working when my health issues became increasingly worse. In 2005, she retired from the VA and we moved fulltime to Florida where we reside today.

LAWLER, CHARLES RICHARD

I joined the US Army in January 1963. After basic, went to Fort Rucker, AL and learned all about Army Aviation. August 1963, I was assigned to my first tour to Vietnam. It was in the Mekong Delta at Vinh Long, with the 544th Trans Det., part of the 114th Air Mobile Company (Knights of the Air). Most of the Helicopters (UH-1As, H-19s, H-21s, H-13, H-23 and H-34s) involved in the war effort at this time were in direct support of US Special Forces in the training and support to South Vietnam Forces. We used to get weekend passes to Saigon and walking alone in the city was not considered hazardous to your health.

Upon returning to the States in 1964, I was assigned to Fort Benning, GA and became part of a new test unit called the 11th Air Assault Division. While assigned to Co. B, 611th Maintenance Bn, we were attached to Co A, 82nd Avn Bn, 82 Airborne Div., sent to California for a trip on the USS Iwo Jima, and thus began my second tour to Vietnam.

In April 1965, landed at Vung Tau, and we were now attached to the 173rd ABN Bde. Later, we moved to their base at Bien Hoa. Stayed with Co. B, 611th as A/C Mechanic until September 1965, in which I was reassigned to the Falcons Platoon as a crewchief until my departure in March 1966. During this period, AC missions dealt with armed protection of Cowboys flights all over the central highlands and the area west of Saigon. Many multiday missions were in support of 173rd troops opening new base camp areas for new arriving US Divisions (1st CAV, 25th, 1st & 4th Inf Divs. etc.) in many areas of the

131

countryside. March 1966 went back to the US—the land of the big PX and left the military.

In the 1970s, I heard it from our President, we were getting out of Vietnam. I rejoined the Army and before I could break in my new combat boots, I was on orders to return to Vietnam—couldn't believe it. I was able to delay it for a while, but eventually Pentagon assigners won and so I left for my third tour.

September 1971 arrived back in Vietnam, assigned to 60th AHC (Ghostriders). The base was upcountry near Tay Ninh and Nha Trang along the coast. By this time, AC mission were significantly different than the previous tours. This third tour seemed dedicated to covering US Forces departing their base camps and withdrawing of US Forces to the coastal areas. In September 1972, my third tour was done and soon after, in 1973, the war was over.

My viewpoint of the Vietnam effort is colored and based on three separate cycles of the war effort in which I was there. In 1963, I thought we were doing a decent job with the local population but was stuck with a corrupt national government. Our Special Forces elements were doing an outstanding job with the local populations, but eventually the national government came into the local areas and screwed up what excellent work was accomplished.

The second cycle, 1965 and onward, we see our own government screwing up the works by backing up the corrupt national government and instead of continuing to force the ARVNs to fight the ground battles, we brought in our own forces to win the battle. During this period, we were blinded to the fact that we were losing the allegiance of the local populations because of our "Search & Destroy mission" and heavy use of carpet bombings and other destructive tactics.

The third cycle, we sort of acknowledge that we couldn't win the battle for the minds of the local population and therefore finally saw the light and worked hard to ensure that we got out of the country with as few causalities as possible.

Overall, the good initial intentions and actions of lower level US forces were overcome by a true lack of understanding by the higher commands (Military and Political) of both the US and Vietnam side of the war effort.

Back in the States, I made Specialist 6, then went to and graduated from Infantry Officer Candidate School (OCS), became an Armor Officer. After many troop assignments, both in the States and Germany, I retired, twelve years later, at the rank of Major. Finished my military career in Maryland and then stayed there awhile and worked as a Facilities Manager with a local MD Federal Credit Union. Retired from that job in 2003.

I am married to Louise and we have two children, who are now married themselves, with two children each. So, I have four grandchildren, two girls and two boys that live in New York and West Virginia. Louise and I moved back to PA (where we were originally both from) after my retirement from my last civilian job. We currently live in a PA 55+ Retirement Community and try to take as many trips of travel to the kids and to take as many vacations as we can.

LAWLER, JOHN W.

Cowboy 19

January 12, 1970, returned stateside and was assigned to Fort Rucker as helicopter instrument instructor, Flight commander, and then Branch commander of IRT section. Instrument Refresher Training for all pilots at Fort Rucker. I attended night school at Embry Riddle Aeronautical University and earned a degree in Aeronautical Science.

Left the Army August 1971.

Hired by Modern American life as VP of sales and marketing. A few years later, I accepted an offer by Merrill Lynch, Pierce, Fenner and Smith as a stock broker. Completed eight months of training and was assigned to their branch office in New Haven, CT. Moved up the corporate ladder and was instrumental in forming Merrill Lynch Life agency which encompassed total financial planning, savings, checking, and insurance services. Promoted to EX VP and transferred to Atlanta, Georgia. I stayed in Georgia for over 28 years.

After leaving active duty, I joined the Army National Guard in CT (Helicopter pilot) then transferred to the Rhode Island Guard. Flew many rescue missions during the blizzard of 1978. Transferred to the Georgia Guard and then California Guard. I retired after 41 years of combined active duty and National Guard. Total flight time 10,129.5 hours.

During my time in Georgia, I was a part time helicopter pilot for the Georgia Forestry commission, fighting forest fires. Also flew for the GBI/DEA, their drug interdiction program in Georgia.

Attend night school for 4.5 years at the John Marshal School of Law, Atlanta, GA.

I left Merrill Lynch after eleven years and started my own financial firm which grew to six additional companies: Investment Advisory, Life Insurance Agency, Property Casualty Agency, Financial planning, Leasing firm (Lease USA), Real Estate Development, which I developed and owned a 40-acre office park. Part owner of Life USA, a re-insurance company. Owned "The Center for Quality Management," trained by William E. Demming, the grandfather of Quality movement who trained the Japanese after World War II.

On the board of directors for E.F. Hutton Life in La Jolla, CA.

Married three times and have three boys and two grandchildren.

Now married for over 15 years to my friend and partner, Georgia, joined her in the real estate business in Southern California. Dale Dilts and his wife were Best man and Maid of Honor at our wedding on the shore of Lake Tahoe. Dale and I were roommates in Vietnam. Dale painted my nose cover for the Christmas Holidays (1969), the picture is published in "Vietnam War Helicopter Art" by John Brennan.

I plan to retire in one year and just be part-time in real estate. We plan to travel 6-8 months out of the year and spend more time with our grandchildren.

How did your service and experiences affect your life?

Evaluated in 2004 as having PTSD. Now I finally understood, years later, why there were problems and frustrations.

I am a member of the California Association of Realtors and the Pasadena Board of Realtors.

LINSTAD, ROLAND (ROLY) EDWARD

I left Vietnam and the Army in June of 1971. Upon arrival to the antiwar sentiment of liberal Massachusetts, I made a command decision. Myself and a friend, that had also just come home from Vietnam, packed up our camping gear and headed to the White Mountains of New Hampshire. For two weeks we would climb a different mountain trail during the day and evenings would be spent in front of a campfire, drinking and telling stories. We never cooked a meal for the two weeks. The other campers took us under their wings and kept us fed. These campers also let my friend and me realize what we had done was appreciated by most people not affiliated with the media.

During the next three years I performed many tasks. I had taken the Civil Service exam to become a Police Officer in 1971 and had been informed that I had been accepted. An injunction had been put on that list, fighting the 5'10" height minimum. Legal battle was over in 1974 with the minimum being reduced to 5'6". My tasks included: auto mechanic, truck driver, and construction equipment operator.

In September of 1974, I began my career as a Police Officer. During my first five years or so, I was a patrol officer that had a side of Tactical Police Force (early SWAT). About this time in my career, I had to decide on taking tests for advancement. I decided against supervisory roles as I preferred working directly with the public. Next duty was Traffic and Safety, including motorcycle unit. After a few years, a Drug Unit (narcotics) was started. I was assigned to the Greater Lowell Drug Task Force for five years. Budget cuts came, and the Task Force was disbanded. Back to patrol for a while.

My final full-time assignment began in 1996 as the Detective in charge of Domestic Violence. This assignment was my most rewarding. My duties were not always bad. At times I assisted couples to see the positive side of their relationship and get back together. Situations that had no future to end peacefully. But most of all I was the voice for the children and made sure they were not used by either side against the other, the best I could. I retired from fulltime police work March 1, 2005.

In November of 1974, I married my first wife, Robin. The marriage ended in 1996, but the result was two boys, Erik (DOB 1976) and Nathan (DOB 1978). They are both very successful in their very different lines of work. Computers and police work. They have given me three grandchildren: Arik, age 13, Kira, age 11 (Erik's) and Nolan, age 9 (Nathan's). I am very proud of my boys and their children.

In 1997, I received a call from my dental hygienist, Bethany. She informed me that her sister had just had a domestic incident in Lowell (next community) and asked if I could investigate. I informed her that I work closely with Lowell and would be there. Months later, Beth asked me to the movies to repay me for my assistance. It has been a little over 20 years later and we still go to the movies and Patriots games together. We were married St Paddy's Day 2001.

Like many of you, I am a member of some military organizations—VHPA, VFW, etc. But the organization that I am most proud of is the Patriot Guard Riders. For those that don't know of the PGR, it is an organization of primarily motorcycle riders that honor veterans in many ways, funerals, arriving back home from overseas, or leaving to go overseas. Never again will service members come home to the reception Vietnam vets received.

My work ethic came from my parents, but the way I work with people came from the 335th.

MILLER JR., PHILLIP CORBETT

I was in Vietnam 16 July 1969 through 4 April 1970. At Bear Cat, I was in the 1st Platoon, "Ramrods," and my call sign was "Cowboy 14." I left Vietnam 4 April 1970, after being wounded 31 March, and spent two weeks in the 106th General Hospital Burn Unit, Camp Kishine, Japan. Sadly, our Crewchief that day, Spec-5 Kimball Sheldon, was killed in action, along with ten Vietnamese soldier passengers. Our Co-Pilot (WO1 Neil Leonard) and Gunner (Spec-4 Kevin Crabtree) were also wounded (burns and shrapnel). Neil returned to the Cowboys to finish his tour.

I spent six months in the hospital at Hunter Army Air Field, Savannah, GA. Kevin went to a different hospital then he and I were assigned to separate units at Fort Bragg, NC. I was at Bragg for 22 months flying Hueys and OH-58s in the 69th Aviation Company, providing general aviation support for the XVIII Airborne Corps Commander and other Fort Bragg units, as well as visiting dignitaries.

My time at Bragg was great! Fort Bragg/Fayetteville was home when I entered the Army...I knew the area and still had friends and family there. One noteworthy assignment was two weeks in Turkey in a UH-1H with auxiliary fuel tanks...no better tourist photo and sight-seeing platform! At Bragg, I knew I wanted to keep flying in the military, so I left the Army September 1972, and entered the University of Florida (Go Gators!) to finish school and get an Air Force commission through ROTC.

School went well, but I did not set academic records...let's just say I got my degree and commission on time! After graduation, I knew there'd be a delay before AF pilot training due to military cutbacks with the wind-down after Vietnam. I had to eat, so I went to work at Disney World as a cash control teller (1974-75). As in any large organization, Disney assigned people just like the Army. I had a History degree; they immediately put me to work in Finance. Go figure! Disney was also good for me in another way...I met a cute redhead. More later...

In June 1975, I entered AF active duty for pilot training at Laughlin AFB, on the US-Mexico border near Del Rio, TX. AF jet flying was different than Army helicopters, but many tricks we learned in Vietnam proved handy... 'instrument flying' and 'formation' are the same no matter what kind of aircraft. I did well enough to remain as a T-38 Instructor Pilot as my first job, but shortly before graduation I was hospitalized with gut pains and permanently grounded from flying. The good news is I stayed on active duty as an Aircraft Maintenance Officer.

After three years at Laughlin (1975-78), my new bride, Beverly, and I went to Ramstein Air Base, Germany...wonderful assignment and four-year honeymoon (1978-82)! Working on F-4Es was demanding, but as we know, airplanes are finicky...the greatest part was working with so many good folks. This was during the Cold War, and we daily felt the presence of the Russian Bear just across the Iron Curtain and East German border. Part of the job required weapons training deployments to Spain and Sardinia...great beaches and superb food! I also luckily got a four-month assignment in 1980 to Cairo, helping train the Egyptian Air Force on 35 F-4Es we "gave" them as part of the Camp David Peace Accords between Israel and Egypt. Living in a luxury hotel just 500 yards from the Great Pyramid was not bad at all...a superb way to serve my nation!

After Ramstein, I went to Randolph AFB, TX, for a four-year staff job at HQ, Air Training Command (1982-86). San Antonio was great, and the job brought more opportunities and added responsibilities. I beat the odds and was promoted to Major, then was selected for Air Command and Staff College, at Maxwell AFB, in Montgomery AL (1986-89). After one year as a student, I was 'invited' to remain on the faculty developing the logistics curriculum and as an in-class instructor. I enjoyed working with stellar folks...my alternate classroom instructor was one of the 52 Americans held hostage in the US Embassy in Iran for 444 days. I still have Paul's hostage bracelet!

Next was the Far East...Kadena AB, Okinawa, Japan, for seven years (1989-1996). That's quite long for military assignments, and I was blessed to command three squadrons (two aircraft maintenance squadrons and the largest munitions storage unit in the Pacific). I was also promoted to Lieutenant Colonel, and enjoyed training deployments to the Philippines, northern Japan, and Thailand. (On our way back to Okinawa from Thailand, I was in the KC-10 cockpit as we flew along the southern tip of Vietnam... strange seeing Vietnam again after 33 years!).

Unfortunately, in 1993, I lost my wife, Beverly, in a scuba diving incident. Obviously, a tough time, but I soon met Linda...love blossomed, and we got married in 1995. God is so good! Before leaving Okinawa, Linda and I visited Vietnam on a 10-day trip...Saigon, Cu Chi, Da Nang, Hanoi, back to Saigon, then down the main road through Tan An and on to My Tho, complete with a Mekong River boat trek and lunch between Dong Tam and Ben Tre. Good memories!

Before leaving Kadena, I was promoted to Colonel, and in 1996, moved to the US fighter base Royal Air Force (RAF) Lakenheath, England, as Logistics Group Commander (1996-98). I was privileged to direct the efforts of 1,776 people in six squadrons providing support to three F-15 fighter squadrons. A personal highlight of living in England was taking my dad and his wife to Normandy. We saw all the invasion sites and stood exactly where 18-year old Corporal Miller parachuted in on D-Day, 6 June 1944. Fantastic!

As fun as Lakenheath was, my United Kingdom tour got even

better with a follow-on assignment to RAF Mildenhall as the Logistics Director, Third Air Force (3AF) (1998-2001). Our responsibilities ranged from Northern Europe down to all sub-Saharan Africa. While at 3AF, I was privileged to take part in two Joint Task Forces (JTF) providing humanitarian relief. First, I was Commander of AF Forces for "JTF Shining Hope" helping refugees fleeing ethnic genocide in Yugoslavia. This JTF happened at the same time as the air war over Yugoslavia, and since we were doing so well in a valuable job, President Clinton, his Secretaries of State and Defense, and the Chairman of the Joint Chiefs visited us at our JTF headquarters in Germany. A year later, I was the Logistics Director for "JTF Atlas Response" helping those affected by massive flooding in seven nations on the southern tip of Africa. Both JTFs allowed our nation to lend helping hands to those adversely affected by horrible life circumstances.

My last AF job was at Southern Illinois University, Carbondale as the AFROTC Detachment Commander (2001-04). I could not have imagined a more rewarding experience than teaching and mentoring young people in pursuit of their degree and an Air Force commission. It taught me a lot about how our nation benefits from the strengths of such dedicated young people. Despite what we may see in the news about "youngsters," our country is in good hands with those who step forward to serve.

I retired from the AF in 2004, after four years in the Army and 29 in the AF. I tried for a position as the Aircraft Maintenance Director with a major contractor supporting the Air Force's flying training base at Columbus AFB, MS. Unfortunately, the company I was with did not get the contract, so Linda and I took that as a sign to relocate to Vancouver, WA. We bought our retirement home and now enjoy living close to Linda's two daughters, their husbands, and all four of our grandkids. Soon after arriving in WA, I started teaching online courses for Columbia College, Missouri, and recently hit ten years of teaching.

What have I done and learned after Vietnam? Life is good, and I'd do it again! As a grateful American, I was blessed to serve my nation for over 33 years, but the highlight of my service is knowing that at one time I was, and always will be, a Cowboy. I was also blessed with the love of two wonderful women. My first wife, Beverly, gave me 15 years of a very good, loving marriage. And the love of my life, Linda, has blessed me with 22 blissful years, fills my heart to overflowing, and as a grandma, she taught me the joys of being a grandpa! But, I saved the best for last, because more important than any earthly accomplishment or achievement, I have the assurance of my eternal salvation because of Jesus Christ.

MOREHEAD, JOHNIE "RAY"

After leaving Vietnam I was stationed in Fort Hood Texas, and six months later we were given the option for an early out of active duty by taking a year on National Guard.

Sheree and I got married shortly after leaving the service (active duty) and have been married for over 45 years, best thing that happened to me after Vietnam. We were married in 1972, had our daughter Nicole in 1976, and in 1998 we adopted our son Ryan. Ryan is a special boy. He is 19 now and been in our home since he was eight days old. He has special needs, he is non-verbal and non-mobile, and requires full time care. We are also raising our eight-year-old granddaughter (for the last couple of years).

We were able after a few years of marriage to send my wife Sheree to college for her master's degree in special education, which she enjoyed very much, and she retired from in 2011 This has also played a great role in raising our son with special needs.

Wasn't happy like all other Vets the way we were treated when we arrived home, they (America) didn't appreciate what we had been through or sacrificed. Had a lot of trouble dealing with that.

After about six years of not being able to hold a job, I got my break in 1977 by using my military aircraft mechanical training and ended up in a large oil field equipment manufacturing machine shop. Worked and trained there for a little over 36 years. Was able to work in a few different facilities of ours in different states and attend a lot of machine and control classes. I ended up being the Senior Tech for the last 25 years. The military discipline and training was a plus for this.

After retiring in 2013 we have just been doing projects around our home and enjoying the American Dream that we fought for.

Vietnam left me with a few problems, PTSD, Type 2 diabetes, and a few more, but we continue to fight.

MULLEN JR., PAUL J.

Cowboys 1966-67

Raised in Upstate New York (Utica), I joined the US Army in August 1964 when I was 17 years old and right out of high school (still a kid). After basic training at Fort Dix, NJ, I went to Fort Rucker, AL for six months of training as a helicopter mechanic. Following that, I was sent to Germany where I worked as a mechanic and crewchief on CH-34 helicopters and then was moved into base operations. In December 1966, I was transferred to Vietnam and joined the Cowboys, first as a Huey mechanic (which I knew nothing about) and then into base operations (which I did). I continued to work in Operations but flying as a door gunner every chance I got. Like most of the guys in the Cowboys, I had to go through tough things that have stayed with me until this day.

I returned home in late August 1967. Again, like many of our Vietnam veterans, I had a lot of difficulty adjusting to civilian life. Finding work, taking orders from non-military people, maintaining a schedule, and just fitting in with a society that treated us like we were terrible people and putting the blame for Vietnam and all its trouble right on our backs. I followed the course of many others and basically crawled into a bottle, making things even worse.

After six months at home, I went on a blind date and met a young woman who I credit with saving my life. Ruth Kittle gave me someone to cling to and helped me get myself straightened out. We will celebrate our 50th anniversary in October 2018. We were married six

145

months after we met and produced two beautiful daughters who have presented us with one grandson and two granddaughters.

After working a series of low-end jobs, in 1976 I went to college and earned a degree in Electronic Engineering. At that time, most electronic technician jobs were in the appliance repair field, but I lucked out and found a job with a large research and development company on an Army contract that was developing ways to "put the computer in the foxhole." This was before PCs, Macs, and other devices that are used today. My main area of expertise was in the computer network area. After several years doing this, I went back to college and received a second degree in Information Systems with a minor in computer programming.

In the mid-1990s, I made a career change in direction and went into technical management, primarily program and project management. My jobs took me all over the country building state of the art computer networks. I was one of the first to incorporate the use of fiber optics in the network field, as well as working on the leading edge of the Year 2000 computer effort. In 1998, and again in 2004, my wife was diagnosed with breast cancer and we continue to deal with all that implies. But she is still with me, so we just do what we must and treat each day as a special gift.

After almost 25 years of continual national and international travel, I decided to wind my career down with a non-travel job and I spent my final work years as a Quality manager for Goodyear Tire and Rubber. In 2011, at age 65 I retired full time. Two years later, my wife also retired. We now spend our time caring for older rescue dogs with health problems, and traveling around the country, but this time in our RV trailer, going wherever the winds blow. We will probably continue to do this until we physically can't do it...or we run out of money...whichever comes first. At that time, we will sit back and watch our videos.

MURPHY, ROBERT BRIAN

After Vietnam in September 1970, I was assigned to Fort Rucker as an instrument IP at Hanchey AHP, flying TH-13T helicopters. For the next two years, I taught Initial Entry Rotary Wing students in the Advanced Instrument phase while I got my fixed wing commercial and instrument ratings on the G.I. Bill. At that time, we were organized under Department of Advanced Flight Training (DAFT). In December 1972, I went to UH-1 IP Course and was reassigned to Lowe AHP to teach Tactics. Our son was born while I was in Vietnam and our daughter came along in 1973. They both became nurses; our daughter right after high school, and our son after a four-year enlistment in the Air Force. Our son lives in Wisconsin and our daughter remains in the Fort Rucker area.

Fort Wolters was moving Primary Flight Training to Fort Rucker, so in early 1974, I and several other Tactics IPs moved to Shell AHP to form VNAF Eval for the Vietnamese Air Force instrument and contact training in the UH-1H. While awaiting a full student load, I managed to get the Fort Rucker Aviation Safety Officer's Course and to challenge the Army Fixed Wing Qualification Course in the T-41.

That job lasted till Saigon fell. In the summer of 1975, I attended TH-55 MOI and then attended the OH-58 IP Course so that I could teach a West Point Private Pilot Course, which occupied my time till I left for Korea.

En route to Korea in Fall, 1975, I attended the AMMEDS Course at Fort Sam Houston, TX. I arrived at the 377th Med Co. Air Ambulance in December 1975. At that time the company was split into de-

tachments at Camp Stanley, Osan AFB, and Taegu. I was assigned to
the North Detachment at Camp Stanley to replace the departing IP.
This was a great assignment till the unit was consolidated at Yongsan,
and the mission was handled as field standbys at Camp Casey, Osan
AFB, and Taegu. So, I lived out of my backpack at these standby sites
till I became the unit SIP, based at Yongsan.

In January 1977, I returned to Fort Rucker, to Cairns AAF to
attend the Rotary Wing Instrument Flight Examiner's Course
(RWIFEC) with a follow-on assignment to teach the Rotary Wing
Aviator Refresher Course till I moved over to teach the Examiner's
Course until I went to the Warrant Officer Advanced Course in 1980.
Following the Advanced Course, I had nine months of Degree Com-
pletion Program to get my Bachelor's Degree in Professional Aero-
nautics from Embry Riddle Aeronautical University.

Next stop was back to Korea in May 1981 and the 2nd Infantry Division, 4/7th CAV at Camp Stanton. I started out in Charlie
Troop then moved to Squadron Standards for the remainder of my
tour. Charlie Troop, Delta Troop, Operations & Standards were all
co-located at Camp Stanton till spring, 1982 when we all moved to
Camp Stanley.

I applied for, and received, an Inter Theater Transfer to Fort Wain-
wright, Alaska, so after a short leave to see relatives in Nevada and
West Virginia, in June 1982 we set off up the ALCAN Highway to
Fort Wainwright and E Troop (Air), 1st CAV. This had to be the best
Cav assignment ever. I spent my first year in the Scout Platoon as an
OH-58 SIP and IFE. I then moved to the Lift Platoon as a UH-1H
SIP and IFE.

In the spring of 1984, I moved next door to be the 222nd AVN
BN Standards UH-1H SIP & IFE till my departure in December,
1984, en route Davison AAF, Fort Belvoir, VA. Assigned to the Military District of Washington (MDW), we had a Helicopter Company comprised of UH-1H/V and UH-60 helicopters, and a Fixed

Wing Company with U-21A, U-21F, U-21J, and C-12A, C-12C, C-12D, and C-12L aircraft. Since I had only a fixed wing single engine MOS, I went to a Helicopter Company. After about a year I got a local multi-engine qualification and moved across the runway to an Airplane Company. I remained in Airplane Company as a VIP pilot and IP/IFE and dedicated PC for Secretary of the Army, John O. Marsh, till Fall 1989 when I left for the Warrant Officer Master Warrant Officer Course en route Yuma Proving Ground.

Laguna AAF had five UH-1s in various configurations for testing to include a NUH-1H with an F-4 alternator to deliver sufficient power to run the palletized GPS and its associated air conditioner. We also had one U-21A, so for the next three years I was the IP/IFE, flying test support as well as post support. During my spare time I obtained my fixed wing Airline Transport Pilot rating. In mid-1992, I received orders for Fort Rucker as my terminal assignment.

I reported to Cairns AAF, 223rd AVN BN and was assigned to the Contract Evaluation section, giving check rides to initial entry instrument students and contractor IPs in the UH-1 and TH-67, and Fixed Wing Multi-Engine Qualification Course students. When the BN Standardization Officer left, I backfilled him as Chief of Standards until I retired in February 1994.

As a side note, my final flight in the Army was with Billie D. Belsher, a fellow Cowboy pilot in 1969/70. We were fielding the TH-67 as the new instrument trainer, and this was Bill's end of IP Course check ride. I viewed it as a very fitting end of a wonderfully exciting military career.

After retiring, I went for my VA physical and discovered that I had a serious mitral valve problem that required immediate surgery. It took me nearly two years to get my Class 2 medical back, so at that point, I decided to forego putting my ATP to work and seek local employment. I took a job with a contractor at Range Control at Fort Rucker. Between three different contractors, I stayed 17 years in various capacities, ranging from heavy equipment trainer to CAD and GIS drafts-man and cartographer/supervisor.

Our time in Fort Rucker saw the birth of our nine grandchildren, and Joyce had the honor and responsibility of home schooling the grands for nine years. In my spare time I raised beef cattle on a small scale, averaging 40-50 Black Angus cows. I also kept my CFI current and continued flying for pleasure and some flight and ground instructing.

In late 2015, we faced the need to acquire some geographical separation from our adult daughter, so we sold our home and pastures and bought a small farm near Montgomery, Alabama. I have allowed my FAA medical and CFI to expire and probably will leave the flying to the younger generation. We are active in our church and are anxiously awaiting the arrival of our first great grandchild. We both spend more time than we should on social media, but at our age, that's ok.

MURRAY, GEORGE W.

After my Vietnam tour (1966-67 as a "slick"), I went to Hunter AAF in Savannah. I left the Army in February 1968 to accept appointment as Special Agent, Federal Bureau of Investigation. This was my dream job since the 7th grade. I served 25 years and retired in 1992.

After FBI retirement, I handled investigations for Georgia–Pacific Corp and Holland Knight Law Firm in Atlanta. In 2006, I opened Murray Investigative Services LLC. MIS is now my retirement hobby job.

I married Tootsie Hickox six months before deployment to Vietnam. Upon return we had three children—Walt, Jon, and Ashley. Now there are seven grandchildren.

Sadly, Tootsie and I divorced in 2010. Both of us remarried several years later and enjoy the grandchildren—the oldest heads off to college next year.

In 2015, I married a widow, Yonnie Wooddall, adding two step-daughters—Emily and Amy, and three more grandbabies. Yonnie is the daughter of a retired Navy Commander. She traveled the world with the military family and graduated from High School in Catania, Sicily. She understands the military influence, but thinks all my FBI buddies are crazy.

Military service definitely instilled discipline and completion of "the mission." Being a veteran is a proud and enjoyable position in life. It is a privilege to render the hand salute during the National Anthem.

Divorce may have been caused a little by Vietnam; however, a long-term undercover assignment probably was a more important factor. I don't expect to answer this question on this side of heaven. My mother always said I returned from Vietnam a different person—from "a good boy" to not such a nice person. I don't believe any of us doing time in Vietnam were not matured beyond our years. I had more responsibility as a Huey driver than I have experienced since.

I am a member of several veteran organizations—-American Legion, Atlanta Vietnam Veterans Business Association, Vietnam Helicopter Pilots Association, Vietnam Crewmembers Association, Mt. Bethel UMC Vets Group.

WELCOME HOME!

OFFRINGA, GREGORY ANDRIAN

(1948-2015)

Submitted by: Barbara Offringa

Vietnam

When I heard about the Post Vietnam Biography Project for the 335th AHC, I thought: "Greg was so proud of his life journey and accomplishments after he served in Vietnam," so I am happy to summarize them and the man he was. You see, Greg, my husband and best friend of 35 years, died from Agent Orange related Pancreatic Cancer in 2015. It pained him deeply to see his fellow soldiers stereotyped as homeless alcoholics, who could not get their lives back on track. To combat that view, he let everyone know that he was a Vietnam Vet, and damn proud of it. Serving in Vietnam made him the man he was and set him on a path to fulfill his life.

Martin Luther King Jr said, "No one really knows why they are alive until they know what they'd die for." Greg said that he would die for the people he loved: his fellow soldiers and later his family. He had faced death many times in Vietnam and eventually adapted to the grim reality that living was not guaranteed. It was dark, but true, and allowed him to focus on serving his country. In the end, Greg was lucky enough to survive Vietnam, but faced a new battle when he was diagnosed with Pancreatic Cancer. He knew that this battle would be different, but his experience facing death in Vietnam allowed him to do the same when he received his diagnosis. That mentality allowed him to live his life, love his family, and do his work.

Greg learned what he'd die for early in his life and maybe that helped him know why he was alive. He was alive for all of us; his friends and family.

Academic Achievements

After serving in Vietnam and a brief stint in his local police department, Greg attended and graduated from Central College in Pella, Iowa as a Sociology major in 1974. He was very interested in different cultures and their world views. This led him to get his MSW (Master of Social Work) from Boston University in 1979 and to practice as a clinical social worker in Boston. But he had another wish. He wanted to do individual psychotherapy and counseling, so he enrolled at the Boston Institute for Psychotherapy ("BIP") for more training. When I met him in 1980, he was working full time at the Jamaica Plain VA Medical Center in Boston as a social worker, as well as going for this training several evenings a week. He treated many veterans at the VA and they trusted him, as a fellow vet who understood combat issues and PTSD. At his funeral, several war veterans came up to me and told me first-hand how he had helped them re-adjust to the civilian world and to face their demons.

He could have stopped there but he did not. He wanted to obtain a PhD in Psychology. By this time, we had two small children and he was working many evenings seeing clients, so he found a program at the Fielding Institute in California that would allow him to do his dissertation remotely. This endeavor did not go smoothly. His advisor disliked him and coincidentally had an extremely negative view of the Vietnam War and the veterans who served there. He dropped out for a while, but then re-enrolled and finished his doctorate with a new advisor in 1997. So, he achieved his dream to be Dr. Gregory A. Offringa!

Personal Achievements

However, Greg's academic accomplishments were just one aspect of the man he was. He was extremely open minded and instinctively cared about others, to a fault. That view of the world allowed us to teach our children to withhold judgment of others. Everyone got the benefit of the doubt—even total strangers—until Greg had time to look deep inside and see what they were made of. It meant that he befriended unexpectedly good people, but also trusted some who didn't deserve it. To Greg, the good experiences always outweighed the bad and he could always see a stranger and open himself up to them. It didn't matter how many negative experiences Greg had, because he was brave.

This description of Greg could lead you to think he made friends by way of convenience, but that wasn't the case. Many times, he actively sought out new people to befriend.

In the 1990s, Greg took advantage of the worldwide web to find lost cousins living in the Netherlands. Even then, the internet was filled with scammers and people who were not to be trusted, but Greg was able to charm and befriend these complete strangers via email. I could imagine the situation playing out very differently, were it not for Greg's approachable friendliness and unrelenting charm.

Then suddenly our family was planning a trip to Holland, where we were greeted with warm hugs from people who didn't know we existed only a few months prior. We stayed in their house where they served us pancakes at breakfast and Dutch gin at night, after the kids went to sleep. All because we shared a last name, an inconsequential amount of genetic material and a few dozen emails.

This success, of course, encouraged the man.

From then on, whenever he saw a chance to make a Dutch connection, he took it.

When we took a trip to Aruba—a former Dutch colony, teeming

with expats—I knew no one was safe. We were traveling with a few other couples and we took bets on how long it would take Greg to tell someone he was Dutch. Some people said a day, one said two days, thinking it would take some time to get settled and might happen at the bar after dinner. A savvy friend said 5 hours, but I bet 45 minutes.

When we got into the taxi 15 minutes after landing, Greg started chatting up the taxi driver and I knew I had already won. "You know, I'm actually Dutch," he said, and I let out a small, victorious laugh. When all the couples met at the hotel, I collected my prize (bragging rights) and informed Greg about the bet. He thought it was extremely funny and it became a running joke throughout the trip.

Greg's love of people was an extension of his bravery. That bravery allowed him to make friends wherever he went; he would talk to anybody, anywhere. On the train to work, at the airport, on vacation, in South Station, the cab on the way to the airport, or the nurses at Dana Farber Cancer Institute while receiving chemotherapy. Sometimes friends from the commuter rail would join us for dinner or the boys who served him pizza at South Station would help him move a refrigerator. It could have been strange, if it wasn't so endearing and it gave him a wide range of friends and acquaintances from an array of backgrounds. If you joined him in a discussion, a poker game, or a scotch, he was your friend for life.

In Defense of Others

Greg was always a big guy, even as a kid, and knew that he was physically more powerful than those around him. In another man, his physical gifts could have been used to push others around, but for my husband, it became a moral obligation to stand up for others. He grew up in rough and tumble Brockton, MA and befriended a small Jewish kid in a town where being Jewish was a rarity. In defending his friend from bullies, Greg knew he would be the sort of man who stands up for those who can't.

In the early 2000s, he learned that Veterans were returning stateside from Iraq and Afghanistan, increasingly with some physical or psychological disability. Greg knew that they were not getting the help they needed and managed to put together a fundraiser for Disabled American Veterans (DAV). He had recently purchased an old police Harley Davidson and rebuilt it into a civilian vehicle, which inspired him to put together a motorcycle fundraiser. He and his friend started in Boston, MA and drove from one Harley dealership to the next, raising money at each location, until the two of them reached the West Coast. The fundraiser garnered a great deal of publicity and managed to raise a little money along the way, after gas and lodging. After the trip, Greg and his friend became even closer than they were before. It was another friendship cemented by defending those who needed it.

Greg left behind an unknowably large number of people who will remember him as a caring friend, dedicated soldier, and savvy psychologist. He had many accomplishments in his life, but I think his biggest accomplishment was how bravely he cared for others, whether they deserved it or not. Greg was a loving, affectionate father and husband and I miss him every day.

OSTERLI, PHILIP (PHIL) PAUL

After Vietnam: In January 1968, my wife, eight-month old son, and I drove across the country to Fort Stewart, GA where I completed IP training and was assigned to the Department of Tactics as assistant flight commander and instructor. In mid-December, we were joined by a daughter, completed our active duty commitment at the end of the year, and began the transition to civilian life by driving our little family back to California. I took a full-time position as a Staff Research Associate on the University of California, Davis Campus where I re-entered graduate school in Plant Genetics and assisted in establishing a Rice Research Facility on campus. Being a Vietnam Veteran on a college campus in 1969 was a unique experience; fortunately, the Davis campus wasn't as vociferous as Berkeley in protesting our involvement in Vietnam, but there was no organized support group on campus either, so I answered a TV advertisement for helicopter pilots in a Sacramento unit of the California National Guard, becoming the first Vietnam Veteran to do so in Northern California. This provided me a unique opportunity to continue a part time career as an Army Aviator as well as a break from the campus environment. In fact, while I was in the process of signing up, that unit was activated for a student occupation of "People's Park" on the Berkeley campus and managed to dispense tear gas from one of their H-19 helicopters over the protesters (and, inadvertently, about a third of the rest of the campus)!

About two years later, I was offered an academic position with UC Cooperative Extension as a Vegetable and Row Crop Farm Advisor in Stanislaus County where I conducted an applied research and education program focusing mainly on dry bean, melon, and tomato production (evaluating new cultivars, irrigation methods, plant nutrition and integrated pest and disease management for the local area) working with commodity associations, pest control advisors/consultants, food processors (dehydrators, freezers and canneries), environmental organizations and regulatory agencies in order to maintain an economically viable agricultural industry in the Central Valley of California. It was truly a very interesting and rewarding job being involved at the interface between cutting edge research and its application in the field. This academic appointment necessitated membership and participation in several professional societies as well as service on various local, regional, and statewide committees. These responsibilities took valuable time commitments in addition to the normal job requirements. Oh, our family added another son and daughter during the early years back in California. Unfortunately, late in this part of my career, my marriage ended in a divorce.

In 1986, our director retired, and several colleagues urged me to consider applying for that position, which, after considering the alternatives, I reluctantly agreed to do so, leading to my being selected by UC to lead our 15-advisor office and serve in the dual capacity as the University administrator and as a county department head—a unique position shared with only 50 others in the state. I was able to maintain some programmatic activity in public policy/land use and was the principal investigator in a unique USDA Water Quality Initiative, developing a new technology that is currently used on over 100,000 acres to reduce sediment loads on the San Joaquin River.

In addition to serving on several state and national committees and organizations, during the last few years on the job I co-chaired the design, construction, and occupation of a new state of the art Agricultural Center that housed all government agencies (county, state, and federal) that work with the agricultural community. During the latter part of this tenure, I fortunately crossed paths with my current spouse and partner, adding two more sons to the family, which has grown to include 13 grandchildren (with one more on the way). Three sons served in the Army overseas in Iraq, Saudi Arabia, and Afghanistan; our oldest grandson served with Special Forces in the Philippines.

That part time military career continued through another 10 years in aviation (including commanding the 126th Air Ambulance Company) followed by six years on the staff of the 175th Medical Brigade, four years as XO of the 352nd Evacuation Hospital, and three as Chief, Personnel and Administration of the 2nd Hospital Center (which convinced me that it was time to transfer to the IRR). I officially retired in 2000 with over 35 years commissioned service. Overall, the military was good to me and my family: I was able to swear in my son as a 2/LT and later joined him in pinning gold bars on his son (my grandson); and I also welcomed a West Point graduate (stepson) and a son-in-law (Air Force) to our extended family.

Throughout my adult life, I always enjoyed good health and led an active outdoor life—boating, water skiing, fishing, and playing adult basketball and senior softball. Unfortunately, I contracted prostate cancer in 1995, and a few years later, colon cancer. While both were successfully treated by surgery, it proved to serve as a "wake-up call," causing me to seriously consider retirement, which I implemented in January 2003, receiving Emeritus status from the University (which explains the email address).

After retiring, I spent a year volunteering at the Great Valley Center (a think tank promoting the Central Valley of California) before deciding it was time to start really enjoying another interest—boating in the Pacific Northwest (cruising in Puget Sound, the San Juan Islands, the Canadian Gulf Islands, the Strait of Georgia and environs), so we moved to Poulsbo, Washington where we currently reside. In addition to the unparalleled year-round boating environment, friends suggested that I try golf, which I still attempt as an excuse to get some exercise (and use as an opportunity to donate a few $'s to these "friends") a couple days a week. About six years ago, the cold, wet winter weather began to affect our aging bones, causing us to become "snowbirds", where we try to enjoy the relative warmth of Central and Southern Coast of California in our motorhome from November to April. Being retired military, we have been able to utilize several military RV parks (gated communities with armed guards) throughout the west coast. Fortunately, many of these bases also have a golf course, which continues to provide an opportunity to chase the little white ball. We recently became active in local chapters of VHPA and DFCS and have long participated in various boating organizations in Washington, serving in several leadership positions. Our extended family is scattered around the country; Texas, Kentucky, Montana, Missouri, and California which affords us an incentive to continue traveling as long as we are able (and the kids put up with us).

In closing, I went to Vietnam because my country asked me to serve. My Vietnam experience was like most veterans: war was the last thing I want to be involved with, but if you're involved you've got to use everything at your disposal; and, you absolutely need to be in it to win… it's not something to be micro-managed from afar! Fortunately, my service with the Falcons/Cowboys made a lasting positive impression on me, providing the necessary training and leadership examples that allowed me to survive in that hostile environment, as well as grow and develop principles and priorities that I hope have made me a more positive, productive member of society and a better husband and parent.

PATCH, DAVID R.

SGM US Army Retired

I served with the 335th Assault Helicopter Company in Vietnam from March 1968 through February 1969. I was further assigned to the 166th Transportation Detachment during this period and served mainly in the maintenance platoon, supporting UH-1B and UH-1D Huey helicopters. This was my first assignment after completing basic and advanced individual training and I generally have favorable memories of this tour.

I flew a few combat missions as a backup crewchief covering other crew members whenever the need should occur, but never got consistently involved as a mission support crew member for any specified period. At that time, the 3rd platoon had "B" model gunships and I can remember a couple of distinct incidents that I experienced on those few missions where I did fill in. One thing I do remember was when I burned my fingers trying to clear a lodged 7.62 round from my M-60 machine gun and never did get it cleared until we got back to the LZ for armament to clear it.

Another thing I remember is when the landing gear was completely torn from a helicopter during a takeoff and, after flying off most of the fuel load to reduce weight, we replaced the landing gear on the helicopter as it hovered above us. I was on the ground assisting as a mechanic during that episode.

I can also remember using a Burgess rechargeable flashlight, spending many nights on a huge PSP (Perforated Steel Planking) flight line performing required aircraft maintenance in preparation for the next day's flight. I can remember doing 25, 50, 75, and 100 hour scheduled maintenance inspections and other maintenance requirements such as shimming collective levers, cleaning barrier filters, replacing broken breakaway door and window safety wires, gearbox flushes, and many other routine tasks as necessary to keep the aircraft mission capable.

I can also remember crewing a supply ship in the company flying up to Qui Nhon or Pleiku for parts runs as needed. That ship was called "Horsethief" and it was also used for recovery operations. I remember working long hours with our Maintenance test pilot, Capt. Grabowski. I served long hours during that time but was rewarded with a very nice R&R in Sidney, Australia at the end of my tour. I have some other faded memories, but the details have escaped me after the fifty plus years between now and then.

After a short leave period at home in Rochester, New Hampshire, in March 1969, I was assigned to an Army Air Field in Baumholder, Germany. There I maintained OH-13H observation helicopters, a U1A Otter, a U6A Beaver, and several O-1 Bird Dogs. This mission was on a huge artillery base in southwestern Germany and the aircraft were primarily used for training, artillery spotting, and personnel transport. We also had one high gloss OD "B" model Huey assigned, but I did not get directly involved with that aircraft very often, because it had an assigned specified crew and was mainly used for VIP transport. I finished my three-year active duty tour there in Baumholder in July 1970 and from there I transferred back into civilian life to my home town.

After a couple of years floating around doing several different jobs around my home town, I pondered reentering the military. I was now married with three children and wanted to stay local to my home town, if possible. I spoke with several recruiters and eventually joined a National Guard unit in Concord, New Hampshire. I was only working scheduled drill weekends at that time but enjoyed the work and being back in uniform. The unit was located just a few miles from where

I lived and, since I was already qualified in a valid aviation MOS, I was easily inducted into the unit without further aviation maintenance training.

Shortly after joining the unit, I was approached by the unit senior maintenance sergeant asking if I was interested in competing for a job as a civil service technician, full time Monday through Friday, supporting the assigned fleet of twelve UH-1V's. I did compete for that position and was selected for the job in the spring of 1973. The unit's mission was primarily air ambulance and patient transport supporting missions within the state of New Hampshire, but also National emergencies when directed by the state's governor. This is a mountainous state where every year inexperienced hikers unprepared for extreme weather conditions or who have sustained serious injuries call for rescue assistance. We also performed search and rescue missions routinely. I was one of several crewchiefs who were often called upon to perform these mountain rescue missions. I stayed at this site performing both fixed and rotary wing maintenance for the next thirty-three years. I eventually attained the position of Aircraft Mechanic Foreman fulltime and retired as a Sergeant Major in the Personnel Directorate in 2005, for a total of thirty-eight years over all.

During my time at the job mentioned above, I was constantly working toward either military or civilian education requirements. After completing each of the military education requirements for promotion to the next grade, I would then work on my civilian education endeavors. I did several civilian industrial arts courses at the Portsmouth Technical Institute, followed by an electronics course at an International Correspondence School in the early nineties. I then earned an Associate Degree in General Studies from Vincennes University in May 1995. After completing the Sergeant's Major academy at Fort Bliss, Texas in June 1997, I continued working toward a Bachelor's Degree. I finished my civilian education with a Bachelor's Degree in Computer Systems Management in June 2002.

After retiring from the military in May of 2005, I began a completely new career working for the state of New Hampshire maintaining computers, printers, fax machines, plotters, cameras, switches, and routers as required for the state of New Hampshire Department of Transportation. I was one of about five technical support members who maintained hundreds of electronic devices located all over the state. After ten plus years performing this work, I retired from the state of New Hampshire.

Looking for a little variety, I then accepted a job delivering car parts to local garages and dealerships. I worked in that capacity for two years and have now started working for the US Post Office part time. I also have been volunteering at a place called the Pittsfield Youth Workshop, counseling kids from several local communities where I was recently recognized as the "Volunteer of the Year for 2017." After a long and prosperous career, I am still enjoying good health and hope to stay productive for many years to come.

PEREZ, JULIAN

Did you work or go back to school? When I was discharged, I used my G.I. Bill to further my education.

What did you go on to do as a career after the war? I became a professional pilot.

Did you get married, divorced, have children? Yes, I did get married more than I should have. However, I did father three great kids... James Carl Perez who went to the dark side (he was a Marine sniper and later worked for Triple Canopy in Iraq for about four years), then Julian Alexander Perez who joined the Army and is with the 20th Special Forces Group, and my daughter Rachel Marie Perez a professional student (just joking).

Did your military experience influence your thinking about war or about the military in general? Yes, very much so. I believe that members of our military are the most selfless individuals that ever graced this earth.... at the same time politicians are the inverse of our soldiers. I abhor war, not only because of the carnage but mostly because our unscrupulous so-called leaders who, in my opinion, waste precious lives for self-gratification.

How did your service and experiences affect your life? It's taken a while, a long while to find my way; I think that I've finally found peace with myself and those around me.

Are you a member of a special organization, club, or association? Yes, the greatest brotherhood on this earth a Vietnam era veteran.

Is there anything you would like to add that we have not covered?

I would like to thank each one of my fellow 335th alumni and particularly my fellow Falcons for giving me a reason to be.

After Vietnam, I attended Texas Tech for a while (Entomology) and went to flight school and finished my pilot's ratings.

After leaving the service, I instructed and flew as a crop duster for nine years then Evergreen Airlines for another nine years and on to United Airline for 27 years.

After retiring, I started training pilots in the Mi-17 (Russian) helicopter for Concord XXI in Daleville, AL and currently I'm an instructor pilot for CAE, training US Army pilots in the C-12 aircraft.

PEYTON, GARY LEE

In November 1967, after a year with the Cowboys as a Falcon, I left Vietnam. I was assigned to Fort Walters and became an IP in the Methods of Instruction section of Flight Evaluation and happily served under CPT Ron Wecas, former platoon leader in the Cowboys. I left the service in October 1969.

After leaving the service, I returned to school along with part time work as a flight instructor for a brief period. That was followed by 50 years of flying for a living, mostly airplanes, but some helicopters.

Until 1976, I was Chief Pilot for the Norand Corporation in Cedar Rapids, Iowa, flying twin Cessna aircraft. During this period, I received my ATP and Learjet type rating.

Until 1999, I flew for the Gates Rubber Company in Denver, Colorado, flying Learjet 35 and 55, Falcon Jets, and Fairchild F27 airplanes and Bell 206 and Eurocopter AS355 helicopters.

After the flight department closed at Gates, I spent two years as a Demonstration Pilot for Bombardier, based at the Learjet facility in Wichita, Kansas, flying Learjet 31, 45, and 60 aircraft.

From January 2001 thru 2007, I was Director of Flight Operations for a division of Verizon in Chicago, Illinois, flying Learjet 60 and Challenger 300 aircraft.

From 2007 thru December 2010 (I turned 65 Christmas day), I flew on several contracts in the Challenger 300/350 aircraft, the longest, more than two years for a family in India.

In early 2010 thru my retirement October 27, 2016, I flew as a contract pilot on the Challenger 300/350 Aircraft for several companies, most notably for Bombardier as an Entry Into Service, Instructor pilot.

I was married during Vietnam, and later divorced and re-married. I raised one son, two step daughters, and a step son. I have been re-married for more than 30 years, and the children all are occupied with successful lives. One son retired from the US Air Force and the other with 22 years in the Army, the girls are happily married with families. We now have seven grandchildren.

After retiring, I settled into doing the things I never had time for. We live in Illinois, north of Chicago most of the year, but have a place in Central Florida for the coldest winter months. Mostly I do projects on the property, our RV, or my latest hobby, my John Deere garden tractor collection.

The thought that keeps everything in order for me with regard to my Vietnam service and experience is, "Those who weren't there, question those of us who were, know there are no answers." Life is and has been good to me.

PINNER, FRANK L.

Submitted by: Karen Pinner

After leaving Vietnam, Frank returned to Georgia Southern University to complete his degree. He graduated in 1971. In 1972, he became engaged to me, Karen Cross. We married on June 17, 1972 in Augusta, Georgia, and moved to Jacksonville, Florida, where Frank worked for the Dept. of Family & Children's Services for several years.

In 1977, Frank took a job in Tehran, Iran working for Bell Helicopter as an auditor. The Iranian Revolution of 1979 cut short this job opportunity! Returning to the US in February 1979, Frank next began working as an auditor with the US Dept. of Education where he remained until he retired in 1997.

In 1985, after 13 years of marriage, we became parents for the first time! A second child soon followed in 1987.

Frank was very active with Christ Lutheran Church, the Georgia National Guard, and in furthering his education. He was working toward a degree in computer programming, but due to his illness, never had an opportunity to complete this goal.

In 1998, Frank was diagnosed with spine and lung cancer, but it was only after his passing that I discovered that the cancer was brought on by exposure to Agent Orange in Vietnam. A life cut way too short.

POSEY, RAY H.

Cowboy 28/3

From Vietnam, to Fort Rucker. Instructor Pilot in tactics. Got caught in RIF in September 1973. Had too much service to give it away. Reverted to enlisted status and was sent to Fort Hood, TX. Assistant S-3 NCO 227th Avn. Bn. 1st Cav. Div. November 1974. The Army, in its infinite wisdom, sent me to Iran. In Iran I was the operations NCO for the ARMISH-MAAG (Military Advisory and Assistance Group) Flight Detachment S joint Army-Air Force fixed wing unit. It was not a MAAG unit but supported MAAG. (Got 75-100 hrs. of U-21 and C-12 time).

September 1976 back to Mother Rucker as an operations NCO at Hanchey AHP. There till 31 August 1979 when I retired.

Knocked around Enterprise till September 1980 when I was hired by Air Logistics flying the oil rigs in the Gulf of Mexico out of Louisiana. Not a bad deal since I worked 14 days on and 14 days off. The pay wasn't all that great, but since I worked six months (less two weeks' vacation) a year, it wasn't too bad.

November 30, 1983 was probably the worst day of my life. My first wife died suddenly and without any warning after 25 years and four days of marriage. Of course, it took time, but I began to heal. Then in 1985, I was introduced to "Super Susan". We were married in October of that year. It's been 32 (working on 33) years.

In early 1990 we bought the first (of three) Motorhomes and began traveling a little bit. We joined a group called S.M.A.R.T. (Special Military Active Recreational Travel Club) a national organization of retired and honorably discharged veterans. The headquarters is in Pensacola and has a web site. If any of you are interested, there are chapters all over the country. We have "Musters" ever couple or three months to socialize and tell war stories.

In November 1987, I was advanced on the retired rolls to my Reserve Rank of CPT. A nice jump in retired pay. By December 2000, I had accumulated 14,000—15,000 hours of flight time. I figured I had pushed my luck about as far as I could, so I retired again and jumped in the Social Security bucket. (Thank you for your support) otherwise, I have become a "yes dear—no dear" household bum.

With Respect to All Cowboys

PREGILLANA JR., ANDRES A.

After Vietnam, I went to a different post back in the States, then on to Hawaii to finish my service. I was discharged in 1970.

After leaving the service, I attended Honolulu Community College to become a carpet, linoleum and soft tile layer journeyman. I was employed with a local company in Honolulu for many years. I married my wife Donna in 1974 and have three children. I have seven grandchildren now who I adore very much.

My military experience had an influence on me as I think back and wonder why I didn't stay in and make a career out of it. I am a lifetime member of the American Legion.

In 1986, we moved to Southern California where I was self-employed in installing flooring. By 1993, we moved again to Washington State where I was employed as a maintenance worker for Johnson Controls, a contractor for the Navy.

In 2005, at the age of 57, I could no longer work as I was totally disabled due to a work injury and the medical illnesses from Agent Orange I was exposed to in Vietnam. I still reside in Silverdale, Washington.

QUATTLEBAUM, JAMES

Two of my most memorable experiences in the Army were the rainy night in January 1968 I arrived at Fort Polk, LA, and the evening in September 1969 when someone came into the Falcon hooch, woke me up and told me I was going home in the morning. While standing on a painted line in the rain at Fort Polk, I remember saying to myself, I got two years of this to endure.

When I was awakened in September in Bear Cat and told I was going home, I really can't recall anything other than a sense of happiness and completion. My transition from Vietnam to civilian life is pretty typical of many draftees. I got on a Freedom Bird at either Tan Son Nhut or Bien Hoa, can't remember which. I arrived at the 24-hour processing center at Travis AFB in the middle of the night. Since I had less than 150 days to ETS, the Army separated me from any further service. I was given a brand-new uniform and money for a plane ticket to Dallas, Texas. I caught a cab in Dallas and went to my father's worksite and borrowed my uncle's car to drive the 30 miles to my hometown of Terrell, Texas. I believe the whole trip took about 72 hours.

My folks were surprised to see me, even though they knew I was close to completing my tour of duty in Vietnam. I believe it was the very next day I put the G.I. Bill to use, I enrolled in Henderson County Junior College, which we all called the Harvard of the Southwest. I completed my third and fourth year of college at East Texas State University. I continued in school and completed my graduate degree at the University of Texas at Arlington. I worked as a psychiatric technician while attending college and after graduation I worked for the Texas Department of Mental Health for about five years. I married my wife Kathy in February 1974 and convinced her to let me go in the Air Force in October 1981. I received a direct commission to the Biomedical Science Corps and 26 years later I retired. My assignments included duty tours in Texas, Guam (three tours), Oklahoma, California, and Alabama (two tours), where I now live.

Kathy and I will be married 44 years come this February 2018. We have three children and four grandchildren. My daughter, Melisa, has three beautiful daughters and lives close by. My oldest son, Ryan, is retired from the Army and has a beautiful daughter. He was a machine gunner and a weapons platoon leader. My youngest son, Kevin, served in the Alabama Army National Guard and currently works and lives nearby. We are fortunate to be able to get together for a meal on the weekends and enjoy the grandchildren.

QUIBERG, OLE REX

I first went into the Army in 1958 and served as a paratrooper with the 82nd Airborne Division. In 1960, I went to Okinawa with the 2/503rd Airborne Battle Group that later became the 173rd Airborne Brigade.

I got out of the Army in 1961 and went to work for the US Forest Service fighting forest fires. While with the forest service, I formed the first Helitack fire crew on the El Dorado National Forest near Lake Tahoe.

I married in 1962. On November 11th (Veteran's Day), 1964, we were in a car accident and my wife died.

In 1965, I decided to go back into the Army. Since I had been out over three years, rank was reduced to a private E-2, and I had to go through basic training again. I got into helicopter flight school and graduated as a warrant officer. Very early in 1967, I was in Vietnam assigned to the Cowboys and supporting my old unit, the 173rd Airborne.

After my time with the Cowboys, I accepted a direct commission to 1LT, Infantry Branch, and stayed in the Army. I got married in 1969.

I was sent back to Vietnam and spent 1971 and 1972 flying slicks with the 101st Airborne Division.

I retired from the Army in 1981 with the rank of Major and went into raising beef cattle in Northern California. In 1983, I went through a divorce, left the ranch, and became a truck driver.

I got married in 1985 and drove a truck in Northern California. That marriage ended in 1987.

After becoming a confirmed bachelor, I married again in 1988. We've been married thirty years. That same year, I started driving long-haul, over the road, in all 48 states and Canada. My wife received her truck driver's license, and we drove as a team for four years. Due to a shoulder injury, she had to quit driving. I then drove solo until 1996 when I retired a second time.

Since 1996, I have rebuilt three houses—two for my step-kids and one for us. My pastime now is doing a lot of reading and a little writing.

I have no children of my own, but I have three step-children, seven grandchildren, and thirteen great-grandchildren.

My twenty years in the Army, including a long time in combat, has shaped my life. It makes me sad to think about all the lives lost and people maimed, physically and mentally, by a war that should never have been fought. I grieve for those, on both sides, for their loses and suffering.

My grandson is married to a beautiful Vietnamese lady who, one month ago, gave us our thirteenth great-grandson. Her father grew up in North Vietnam. I have spent time talking with him and sharing experiences.

I am a lifetime member of the Vietnam Veterans of America, the Disabled American Veterans, and the Veterans of Foreign Wars, but due to a severe hearing loss am no longer able to attend the meetings.

Finally, I cannot put into words how proud I am for having served with all the great guys in the 335th Assault Helicopter Company. Everybody performed above and beyond and did one hell of a job! Thank you for the experience.

RASCH JR., MARTIN E.

Well Dom, I will try to keep this short, but it spans 26 years of service. I started out in headquarters and headquarters Company, 303 Armor, 41st Infantry Division in Centralia, Washington Army National Guard. From there I went to basic training at Fort Ord, California for eight weeks at a beautiful training facility by the seaside. After basic I went to Fort Knox, Kentucky for four weeks of armor training in a tank. I finished up the rest of my six months of active duty at Fort Knox in the same company doing this and that.

I returned home to headquarters and headquarters Company, 303 Armor, 41st Infantry Division in March. I reenlisted in the regular Army for three years and got a duty station at Fort Lewis, Washington, which is only 50 miles from my home in Centralia. There I was assigned to the 34th Armored Battalion, Fort Lewis, Washington.

In 1964, I was PCS overseas to Headquarters and Headquarters Company 19th Infantry, Seventh Army in Asperger Germany. It was nice there, but I didn't like the infantry, so I reenlisted to go back to an armor unit in Munich, Germany. There I was assigned to the 34th Armored Battalion, 24th Infantry Division. From there, I volunteered for Vietnam, as there was a letter that come out stating they needed troops for Vietnam. So, I thought I would go as my father was there at the time. Then I got a big letdown because my company commander and XO decided I was not good enough. So, I wrote a letter to the Secretary of Defense Robert McNamara, he in turn give it to General Johnson, the Chief of Staff. My company commander was ordered to get me cleared within one day. On my way back to the States on my way out of the company, I was asked by the company commander to write a letter that would soften the blow on him because of what he did to me. He was blackballed from making rank any higher and he had to get out of the service.

I spent 30 days at home and in May 1966 I was supposed to go to the 52nd Aviation Company in Pleiku, but instead of that assignment I was sent to the 173rd Airborne Brigade. I was assigned to an Engineer Company.

They paid me $55 extra a month for jump pay and I told them I was not authorized this because I was not jump qualified. So, at the end of six months after doing numerous field training exercises in search of the enemy, the company commander, Capt. Manwille said that I could go to jump school. I told him no thanks because I feared heights.

So that is when they transferred me to A Company, 82nd Aviation, which was later changed to the 335th Assault Helicopter Company, as a cook. During my time there, I met a lot of great people and I thought very highly of the company commander. I respected all the officers and warrant officers in the company. I got to know most of them when I was the cook in the officers' mess and a lot of those people loved fried bologna and fried spam.

There was a SPC 6 that ran the officers mess in the company club and the officers club. He got me involved with working with him in the club, going to Saigon picking up beer and everything else for them. That was great fun. There was a time when I was the night baker and provided night chow for the 166th Transportation people working at night. I enjoyed working with them and feeding them. At the end of my year in country, I extended for six months to become a door gunner on a UH-1, as the First Sgt. said, "No problem, we will have a slot for you." So, I went home for 30 days and when I came back, the First Sgt. told me, "We don't have a slot for you, so we are going to put you in supply." I worked in supply awhile and then I was moved over to tech supply. There I was able to go to Qui Nhon to put all my cards in to order parts for the helicopters and thereby get the job done a lot faster.

We then made a move from Pleiku to a place called Dak To. Sometimes I worked day shift and sometimes I worked nights. In December 1967, just before the Tet Offensive, I rotated back to the States to Fort Leonard Wood, Missouri to F 2-1, which was an AIT training brigade for engineers. Halfway through that assignment, I come down on orders to go to Korea to the Second Infantry Division. My last six months I spent up on the DMZ with a company called ACTA. For those six months, I drew combat pay.

When I left Korea, I went back to Fort Leonard Wood, Missouri and decided I was not going to be a cook anymore. I wanted to go to Drill Sgt. School. After six weeks there, I graduated and became a Drill Sgt. I was on Drill Sgt. status for two years. They tried to send me back to Vietnam after my first year on troop trail and I said no, I'm guaranteed two years here in the States before going back. So, after a while I was assigned to go to a red eye gunner course down in Texas. That was a three-week gunner course and then I was assigned to Germany to the 77th Armor. This time in Kirsch Gose and there we did numerous exercises up in the Fulda Gap.

In the middle of my assignment there, my wife decided she wanted a divorce, as my unit spent so much time in the field. So, she left Europe. The next morning, I caught a flight to McCord Air Force Base in Washington and filed my divorce. I was supposed to go back to Germany after 30 days but ended up getting reassigned to Fort Lewis to the 77th Armor Battalion as a red eye gunner Sgt. and then I got back into tanks.

When I could not get promoted to E-6, I got put out under QNB in July 1976. I went down to my home guard unit where I first started out. Within one week, I was promoted to Staff Sgt. and had my own platoon to run.

Somewhere in that timeframe, my father had a major heart attack in Iowa, so I went back home. I then got into a Drill Sgt. unit and was a senior Drill Sgt. in the company. But they were screwing around, and I couldn't get promoted so I went back to an Armor unit up in Fremont, Nebraska. There I got screwed even worse, so I decided to end my 26 ½ years of military service and retire as a Staff Sgt.

I think this about sums up my 26 1/2 years of service.

ROHLFSEN, ROBERT K.

Ramrod 15 1969-70

After Vietnam I was assigned to a Cav. unit in Fort Carson, Colorado. It just wasn't the same as being with the Cowboys. After five months, we were given a chance for an early out. Many of us started civilian life.

After leaving the service, I became an accountant for a local Grain Cooperative in Iowa. I stayed in the grain and agronomy business until retirement. I started as an accountant and ended up doing everything imaginable in the agronomy business: counter sales, driving trucks, spraying crops, dumping grain, managing a grain cooperative, fixing grain elevators, and grain merchandising. I have served as an EMT and fireman in our local volunteer fire and rescue departments for several years.

After retiring, I remain very active in our church and have done everything except serving in the pulpit. I also am the current Quartermaster in our local VFW and the Adjutant in the American Legion. In the last ten years, I have had the privilege to serve in the Honor Guard for over 100 of our local veterans. I was also a volunteer driver for our Veterans Affairs office, driving veterans to the local VA clinics and hospitals.

After Vietnam, I also met and married my wife, Gloria, and we have enjoyed more than 46 years of being together and raising three girls. Now that they are married with families, we spend a great deal of time traveling to see our kids and grandkids. Jeri and Ryan Linneman

(Elise and Luke) of St. Michael, Minnesota. Leah and Patrick Mc-Manus (Beatrice and Simon) of Canton, NY. Gretchen and Michael Greaves (Jackson) of Auckland, New Zealand.

Every two years we enjoy very much getting together with the guys at the Cowboy reunion and look forward to doing the same in 2019.

SCHIERS, THOMAS C.

1967 – 1968

After returning from Vietnam in February 1968, I was assigned to Hunter Army Air Base in Savannah, Georgia to complete my remaining 18 month obligation in the Army. My duty assignment was a maintenance platoon lead. We maintained helicopters used for pilot training.

While in Savannah, I had two very important life changing events. First, I married the love of my life, Becky. We will be celebrating our 50th anniversary August 9, 2018. Second, I started taking classes at Savannah State College to recover from my less than stellar performance in high school. Using the G.I. Bill, I went on to graduate from Idaho State University with a degree in Electronics Technology and ten years later completed my second degree, a BS in Business.

Becky and I have two grown children, a son Rob and a daughter Heather. We have five grandchildren and two great grandchildren. All reside near us, so we have been very fortunate to be a part of their lives from their births to present. We have resided in Idaho since leaving the service. Thirty-five years in Pocatello, Idaho and the last fifteen in Boise, which we have really enjoyed.

My main career was thirty years in the semiconductor industry. I started as an entry level designer, working through various technical and business assignments and ended with the last seven years as the Vice President of Business Operations. After leaving the semiconductor business, I have been consulting. Working mostly with high tech and home building companies and help manage some large trusts, taking care of business and real estate interests. This is my idea of retirement.

We have been very fortunate to have done a lot of world travel during my career and continue to this day. The last fifteen years we have traveled extensively with our family group of close friends. We especially enjoy going back to other countries where I had business during my career, visiting the friends and acquaintances. My wife, Becky, formed even more friendships than I because of our business relationships and travels. She does an excellent job of staying in touch and occasionally puts together a trip to visit.

My personal interests have been running, (which has now become more hiking), white water rafting, boating/canoeing, travel, wood working, fixing up old cars/trucks, week-ends with Becky just getting away, etc.

In recent years, I have been fortunate to have my grandkids on hikes and white-water rafting trips with me. One special hike we have done a few times, called Hurricane Pass Hike, is from Driggs, Idaho, over the Teton mountain range, to Jackson, Wyoming. It is 22 miles and takes us about nine hours. A long day! Another very special activity was this last summer I took my kids and four of the grandkids on a seven-day 105-mile white water raft trip down the Middle Fork of the Salmon River. I have run my raft on this river over 25 times and wanted to take my grandkids on their first trip down this special river.

I was 18 years old when I joined the Army and 21 when I left active duty. My Grandfather was in the Army during World War I. My Father was in the Army during World War II. Joining the military to do my duty seemed the right thing to do. Because of my experience, I learned a lot about myself and others very quickly. One of the main things I learned is the value of commitment to yourself and others. If you say you will do something, do it. Not doing so in the Military can quickly cost lives. Another valuable lesson is that most people are good. The life-long bonds and friendships formed during stressful combat situations become very rewarding and comforting as the years pass. Being an active member of the American Legion and staying involved with other 335th A.H.C. veterans, that Dominic Fino and others have done an excellent job of keeping together over the years, helps keep my military experience alive and in proper perspective.

A lesson, I believe many of us have learned, and continue to be reminded of, is to question the competence and motives of elected and appointed government officials. If their actions support our country's security and Constitution, they are worthy of their positions. If not, get them out. Too many lives, on both sides of military conflicts, have and continue to be lost due to poor decisions by incompetent government leaders, most having never served in the military.

I have had, and hopefully will continue having, a good life and much of this I attribute to the life perspectives I got from my military experience. A thought I often have is about the many companions that did not come home. They also had loving families, friends, interests, futures, etc. Their futures were cut short and mine was not. Why, is a thought I always carry with me.

There will always be a special place in my heart for my experiences and life-long friendships formed through the 335th A.H.C. and particularly the Falcons!

SCHIMPF, MARK T.

I returned from Vietnam September 1970 and was assigned as a helicopter gunnery instructor at Hunter/Stewart AAF. While awaiting a slot for Cobra IP school, I was offered an early out and took advantage of that opportunity.

I went back to college and completed my Bachelor of Arts degree. I worked for the FAA for seven and a half years but really missed flying. In 1982, I became a law enforcement pilot with the Florida Marine Patrol, flying Hughes 500 D/E models and a Beech Baron. I worked in the Miami/Fort Lauderdale area, pursuing the infamous Cocaine Cowboys during the height of the Columbian drug days.

In 1985, I took an opportunity to work for the St. Lucie County Sheriff's Office. They had a twin-engine airplane but no pilot. The Sheriff also wanted to start a helicopter EMS program in the county. I spent the next thirty-one years flying a variety of EMS helicopters and fixed wing aircraft. I became the Chief Pilot, eventually becoming a Division Commander in all of divisions of the sheriff's office.

Through my law enforcement and military contacts, I was able to procure a retired UH-1B gunship for display at the National Navy UDT-SEAL Museum in Fort Pierce, FL. I retired January 2017 with 34 years in law enforcement. My logbooks show I have flown 109 distinct types of aircraft over the last 54 years.

In 1990, I met the love of my life, Erin. We were married in 1993 and were fortunate enough to raise two fine sons, Colin and Scott. Both boys are excellent students. Colin is attending the University of Miami, majoring in Accounting and Finance. He will pursue his

CPA. Scott is in high school and is deep into computers and coding. He is exploring the field of Cyber Security through summer camps, including some NSA-sponsored ones.

My combat experiences definitely influenced my view of the world. I was faced with my share of flying emergencies over the years and was blessed with the ability to handle them all successfully. People say I never get flustered, remaining calm and collected no matter what high intensity event is taking place. I attribute my calm demeanor to my exposure to some pretty intense situations during my combat tour in Vietnam.

Forty-eight years have passed since I returned from Vietnam. I had many unanswered questions when I got home but due to the unpopularity of the war, I let things lie where they were. As the years passed, my curiosity about the Vietnam war returned and I began researching everything I could find, trying to make some sense of the whole thing. Arguments could be made on both sides as to whether we should or should not have gone to war in Vietnam.

The one undeniable conclusion regarding the war is that it was not fought correctly. Instructions on how to fight the war were given by politicians and their advisors half a world away. We operated with "Rules of Engagement" that restricted our ability to be successful. The other side fought with no rules and we were bound to so many senseless rules that winning was not even a remote possibility. Had control of the war been given to the field commanders and the restrictions lifted, I believe we could have easily won in Vietnam. But as we all know, that didn't happen, and the rest is history.

I feel good about my decision to join the Army and fly helicopters in Vietnam. I learned many things at the ripe old age of 21 that have stayed with me my entire life. I wish we would have been treated better when we came home, but it was a different time. I tip my hat to all Vietnam veterans and in particular those who saw combat.

I have been a member of the Vietnam Helicopter Pilots Association since the early 1980s. Staying in touch with old Army friends really helps heal old wounds. A friend acquired an "H" model Huey and I have been able to fly it on many occasions. Wonderful old helicopters that are a joy to fly. I'm a lucky man!

SCHNEIDER, TIMOTHY

I arrived in Vietnam February 28, 1970. My MOS training was as an O5B Radio Operator and O5C Radio Teletype Operator. On arrival at Bear Cat the first week in March, I was assigned to OJT Avionics Mechanic. I worked on that job until the move to Dong Tam where I became the Acting Communications Sergeant and established ground communications for the Cowboys and the Emu's.

Scrounging equipment became our life. I returned to the States on February 21, 1971, and was discharged from the military.

Three days later, I returned to Ferris State University in Big Rapids, Michigan. I met my wife Kathy while at Ferris and we married in July of 1972. We lived in Flint, Ann Arbor, and Ypsilanti where I worked with Sergeant Van Dudley who was my replacement when I left Vietnam. I worked as a greens keeper, as a construction worker, and as a retail sales manager.

After moving to my present location in the Mt. Pleasant, MI area, I worked in industry for 22 years and then worked for the State of MI Dept. of Corrections until my retirement in 2012. I have one son, Matthew (USMC combat veteran Afghanistan and Iraq), daughter-in-law Monica, and three granddaughters, Kelly, Kenzie and Ava.

At age 70, I am in good health and continue to enjoy hunting, fishing, camping and gardening. I also attend the Vietnam Veterans Reunion each year in Kokomo, Indiana. Life is good!

SCHULTZ, KURT EDWARD

After I returned from Vietnam in September 1966, I still had 18 months to serve on my second enlistment. Somehow the Army screwed up and sent me to my home State, Colorado. Almost too good to be true.

I was stationed at Fort Carson, Colorado with the mission of creating and training Assault Helicopter Companies for deployment to Vietnam. During that time, we successfully deployed two companies to Vietnam. My duties involved arming UH-1C Gunships and training the personnel on the armament systems.

Upon separating from the Army in January of 1968, and after a three month break, I returned to Fort Carson and trained another Assault Helicopter Company for deployment, but this time as a civilian. Since this was the last Assault Helicopter Company that would be deployed from Fort Carson, this job was over, and it was time to move on to something else.

Before I entered the Army, I had attended two years of College at Trinidad Junior College in Trinidad, Colorado. Therefore, I decided to use my G.I. Bill Benefits and continue my education at Colorado State University. After one semester, it was evident to me that I was not college material anymore and needed to get on with my life.

Since I was married and had one child, it was imperative that I get into work that I liked and that had opportunity. After working for a brief period in the sporting goods industry and then managing a guest ranch for two seasons, I decided to start my own guide and outfitting business. This was a natural for me since my father worked for the US

191

Forest Service and I grew up on Ranger Stations and in the backcountry of California and Colorado, developing a love for the outdoors, wildlife, hunting, and fishing.

I started K.E. Schultz Guide and Outfitting Service in 1970, which encompassed the Frying Pan River Valley of Colorado and the adjacent Holy Cross Wilderness, the Hunter-Frying Pan Wilderness, and the Mount Massive Wilderness Areas within the White River and San Isabel National Forests. The business conducted services for wilderness horseback trips, fly fishing and big game hunting to include black bear, elk, mule deer, bighorn sheep, and mountain goat. We also provided backcountry and wilderness pack services for various State and Federal Agencies, plus private entities.

In 1984, we expanded the business and formed an exclusive fly fishing guide service which provided guided trips on the Frying Pan River, Roaring Fork River, Crystal River, and Colorado River by means of walk/wade and float trips by McKenzie Drift Boats. These services were provided through Fothergills's Fly Shop in Aspen, Colorado and Frying Pan Anglers in Basalt, Colorado.

We successfully operated the Guide & Outfitting Service through 1997 when we sold the business. During the period of 1970—1997, I was actively involved in the political side of the Guide & Outfitting industry. I served as President of the Colorado Outfitters Association in 1975 and 1991. I was appointed to the State of Colorado Outfitters Licensing Board in July 1983 thru January 1987 and served as Chairperson July 1983 thru August 1985. I served on the Advisory Board to the Office of Outfitters Registration July 1988 thru December 1991.

After the sale of the Guide and Outfitting Service, I started a new business called The Colorado Horse and Mule Company. This company bred, trained, and sold quality Quarter Horses and Mules. It also provided horse and mule pack seminars, contract packing services, and specialty packing equipment. We operated this business through

2008, after which I retired to become a fly fishing addict, enjoying many days fishing the western rivers of the United States.

My wife Leslie and I have been married since August 1964 and have successfully raised two sons and a daughter. Our older son completed a career in the US Army and retired as a Chief Warrant Officer Four/Apache Helicopter Pilot. He is currently employed by Boeing as a Production Test Pilot on the AH-64 and AH-6 Helicopters. Our younger son initially served with the US Army as an Infantryman with the 82nd Airborne Division, subsequently received an appointment to West Point Military Academy and received a commission as an Infantry Officer and served his commitment with the 82nd Airborne Division. Upon completion of his military service, he attended Harvard Business School and received his MBA. He currently is employed in the corporate sector. Our daughter is employed in the Internet industry as an IT and owns and operates an e-commerce store. We have six grandchildren and three great grandchildren.

My service in the Army and Vietnam greatly influenced my decision to work in the outdoors, which allowed me to raise my family in an environment that would prepare them for whatever endeavor they may choose. My military experience was helpful in operating a successful business through the good times and bad. Without that experience, I don't think I would have been successful.

I am proud to be a Vietnam Veteran of A Company, 82nd Aviation Battalion, 173rd Airborne Brigade (Sep). If I had to do it all over again, I would without question.

SESTER, EDWARD D. (DON)

Served in the Cowboys from 08/70 thru 09/71

Attended and graduated from Embry-Riddle Aeronautical University with a Bachelor of Science Degree in Aeronautical Science in 1975. Obtained Commercial Airplane Single Engine Land, Instrument Airplane/Rotorcraft-Helicopter Rating and an Instructor Rating for Fixed Wing. I did Flight Instruction until I realized there was no money in it.

I used my Minor in Management in various fields until I first went to work for the Federal Government at the United States Department of Agriculture in New Orleans, LA. I ended my career with several years at the Internal Revenue Service in 2013.

I have lived all over the country, from Daytona Beach, FL to Oklahoma City, OK to Tulsa, OK to New Orleans, LA to Kansas City, MO and finally settled in Cape Girardeau, Missouri.

I met the love of my life with my second wife Jo Ann and have been happily married for almost 20 years. I have a beautiful step-daughter and a rambunctious two-year-old grandson. I have a beautiful 18-year-old granddaughter I have been blessed to recently reconnect with and a very smart 11-year-old grandson who lives in Panama City, FL.

SIMONE, LARRY

I was in the 335th from 1970-1971. On a day off, about to DER-OS, I was getting some sun, so I could go home with a "tan." A CPT throws me a set of orders...orders said APO NY. I said where is that? He said, "Germany." I said, "don't we get stationed in the States after Vietnam?"

I went to Germany but on leave I got married. Once there I decided to re-enlist after I saw they were giving a $10,000 bonus. I then went to the 101st at Campbell and back to Germany. I then received orders for Corpus Christi Army Depot. I could not believe I was assigned to a depot. While at the depot, I put in for a Walking Avn Maintenance Warrant and went from SFC to CW2.

First officer assignment was in Korea. We were fielding the first Hawks. From Korea, I went to Fort Hood and interviewed for the new Apache Training Brigade about to field 26 units to the Army. I then retired and returned to the depot as a civilian for 21 more years.

I loved the depot, especially bringing on the Blackhawk Apache and new Chinook. I rose to be the director of aircraft production with 1,000 civilians under me. So, all total I spent 42 years in aviation. Still miss and love it.

SMITH, GALE N.

After Vietnam and went to the Armor Officer Advanced Course at Fort Knox, Kentucky.

After the war, I remained in the Army for a total of 29 years. I then worked in the private sector for another 21 years. While in the Army, I moved 34 times in 29 years, three overseas tours, participated in Desert Shield/Desert Storm. After the Army, I move to Virginia and have lived here since. I jokingly say my next move is to Arlington (National Cemetery) if there are any slots when the Lord calls me home. I participated in Desert Shield/Desert Storm as one of nine strategic planners that developed the plan to liberate Kuwait. Our plan was approved by Secretary Chaney and General Powell the week between Christmas and New Year, 1990. As we all know, the operation began in January 1991 and was an overwhelming success. Significantly, we planners strongly recommended not invading Iraq unless we had an end-state that would be supported by the impressive coalition of forces that included Syrian and other forces that were not necessarily friendly to the United States. We used simulations to rehearse the operation the first week of January 1991 and our Order of Battle assessment (Intelligence Preparation of the Battle Field was the term at the time) of how Iraq would defend was 98.8% correct. The result of a quick victory was the evidence of years of training our senior generals and officers to understand the Operational Art of War. That said, it was the Noncommissioned Officers and Soldiers that made it all happen, as is always the case. The late Speaker of the House, Tip O'Neal, once said, 'All politics is local.' To paraphrase, 'All battles are local and resolved by those on the ground…our junior officers with their exceptional NCOs and Soldiers.'

Did you get married, divorced, have children?

I was married in December 1967 before I went to Vietnam in September 1968. Connie and I are still married and will celebrate our 50th anniversary this year, 2017. We have three sons and two grandsons. One lives in Virginia; one lives in Atlanta; and, one lives in Denver. We get to see them all frequently and certainly at holidays. They are all more and more interested in what happened in Vietnam and recall their youth as Army brats. One grandson lives in Omaha and the youngest lives in Denver. Since retirement from all 'paid' work in 2015, we travel a lot. We spend winters in Puerto Vallarta, Nayarit, Mexico; spring and fall in Virginia; and, summers in Virginia. At least that's what we've done thus far. We enjoy the Cowboy gatherings and learn more each time we are with the group.

Did your military experience influence your thinking about war or about the military in general?

I was fortunate to spend the majority of my first 20 years in command positions. I commanded five company size units to include Armored, Cavalry (Air & Ground), and a two-year command of A Squadron, The Blues and Royals, Household Cavalry, British Army; and two Armored Cavalry Squadrons (1-10 Cavalry and 2-7 Cavalry). I attended the Joint Staff College (1979) and the Army War College (Class of 1988). We had two tours in Germany, one tour in England, and three tours at Fort Carson, Colorado (Air & Ground). I commanded a ROTC detachment for four years in Texas and taught Texas & American government in addition to the military courses. Along the way, I earned a Master's degree in Public Administration to complement my undergraduate degrees in Political Philosophy and International Relations.

Both my life experience and education taught me a lot about war and the consequences. The biggest lesson, however, is the importance of an informed electorate to vet crony politicians with power to put Americans in harm's way. The challenge is to get citizens to understand the extensive power of the President as Commander in Chief. As a student at the Army War College, we studied extensively the various philosophies of War, the causes, and the outcomes. All of this was brought to life walking the US Civil War and European battle fields where so many died needlessly.

How did your service and experiences affect your life?

It had a profound impact. My last nine years was as a senior officer where I was involved in setting up an Army program to train two, three, and four-star generals on how to learn and understand the Operational Art of War. It was called Battle Command Training Program (now Mission Command Training Program). It was a two-part event that started within the first six months of a general taking command at the various levels of command. Part I was a week-long seminar where we immersed the commander and his staff in a scenario that matched their primary war mission (primarily the Soviet Union and Korea). The purpose was to help the commander and his staff learn to use their SOPS, Army Doctrine, and learn lessons from past operations. It had an active Opposing Force that had to do its own mission planning using Soviet or North Korean tactics and doctrine with the intent of providing an extensive, relentless, challenging operation for the units to practice their mission tasks.

Part II was a computer driven, 5-7-day exercise 24/7 with a pace that was ruthlessly relentless. The length was chosen so that those being exercised and trained had to develop sleep plans and work as a team. The realism was beyond comparison to battle. Leaders that had been in battle often told us that the experience and uninterrupted intensity was like nothing they had experienced in battle. The results were demonstrated in Desert Storm freeing Kuwait and in the initial Iraq invasion where skills were manifest near flawlessly. The result of both 'victories' were lost by crony politicians, just as they did in Vietnam. The Soldiers performed brilliantly but were failed by their civilian 'leadership.'

After doing BCTP for 39 months, I became the first Director of the Army's National Simulation Center at Fort Leavenworth, KS. As Director, I pioneered the use of large scale virtual and constructive simulations to practice fighting two major theater of was scenarios near simultaneously. In that capacity, we developed a methodology where each military service simulation was able to work together to replicate their mission needs. To illustrate, a bomb dropped by a fighter from the Navy or Air Force model would land in the Army's model where battle damage assessment could be done. Air Defense systems in the Army and Navy models could engage and destroy planes in all models. I worked with several National Laboratories to get this done, including The Jet Propulsion Lab, Oak Ridge National Labs, Las Alamos National Lab, and Johns Hopkins National Lab.

My last assignment in the Army was with the Office of the Chief of Staff working on a Task Force. It was purposely modeled after one created by General Leslie McNair before the Second World War called The Louisiana Maneuvers Task Force. The purpose was to help the Army implement a Base Realignment and Closure (BRAC) mandate and downsize after Desert Storm from a 780,000-active duty force to a 480,000-active force within three years. This was another example where crony politicians misread history.

Unfortunately, they believed that once the Berlin Wall came down and the Soviet Union collapsed, there was no reason to retain a large standing Army. Sadly, the missions didn't change but the funding did. The result is that when the attacks on New York and the Pentagon occurred in September 2001, once again the Army was not prepared to do its mission to defend the nation. The 'peace dividend' had taken its toll because the modernization had not occurred, the training funds were near non-existent, and repair parts were scarce to none. Sadly, when the Iraq invasion began in 2003, there were no or few rotor blades, no or few tracks for vehicles, and the nation again had to modernize while at war. We all remember the horror stories of Moms doing bake sales to buy and send armor plates to their Soldier sons and daughters.

Are you a member of a special organization, club, or association?

I am a life member of the Armor Association, the Army Aviation Association, the Association of the United States Army, the Army War College Association, the Military Officers Association, the Armed Forces Mutual Aid Association, and several civic organizations. I dropped my membership in The VFW and American Legion because I didn't like the way they were being operated.

Is there anything you would like to add that we have not covered?

After I left the Army in 1994, I joined a consulting company and spent 21 years working with DoD and predominantly the Army and the Combatant Commands in the Pacific, Europe, and Africa. Sadly, I watched the devastation caused by sequestration and the toll of numerous tours into war zones on Soldiers and families. It seems to me that America is incapable of learning any lessons from War and so quickly forgets. I was involved in a joint effort with DoD and the Veterans Administration to streamline and simplify the transition from the service systems to the VA. Sadly as is so often the case, the implementation was flawed and got bogged down in bureaucratic pettiness around who paid and when. The problems continue and are worse today. I read an article recently that says there are now as many veterans from the current wars in Iraq, Afghanistan, and other operational areas that matches those of us from Vietnam. Too many citizens don't know or forget that the first role of government is to protect we the people. Too frequently it fails and needless harm to Soldiers and family is always the result.

SMITH, MILTON A.

Cowboy October 1967 to October 1968

Wow, not sure where to start. After Vietnam, I still had 1 1/2 years left on my enlistment. I was 17 when enlisted and 18 years and four months old when I arrived in country on October 14, 1967.

After Vietnam I was stationed at Fort Eustis VA and as an E-5 was an instructor at the 37N20 School. I did that until I was discharged in April 1970.

I was really surprised and upset at how people treated me upon my return. I enlisted in the active reserve upon discharge and served for another seven years as a scout platoon staff sergeant for an armor company. During that time, I worked at Bethlehem Steel for 18 years and went through the millwright apprentice program and became a certified welder. I also started a Mechanical Contracting business and have been working in that for the last 30+ years. We have about 50 employees.

I drank heavily when I returned from Vietnam and that continued for about eight years. I have been married three times and divorced twice. I have eight children, seven boys and one girl.

As far as my thinking about my service in Vietnam, I'm very proud that I served, and we did an awesome job! Go Army! Go Cowboys!

I was very affected and still am by the Vietnam War. I suffered and still do from PTSD and had a tough time until recently realizing how much the war had affected and changed my life. I was abusing alcohol for a long time, did extreme physical activities, and became a worka-

201

holic. I also learned to put up walls in my military and civilian life, so I did not get too close to anyone anymore. Basically, I lived on the edge.

I have not retired and have no plans to because I would have too much idle time and think about things too much.

I cannot say enough about the men I served with in Vietnam who were absolutely amazing, brave, caring and they kicked ass. Did our job and were never anything but honorable.

S.SGT Milton Smith

SNYDER, DONALD L.

I was discharged September of 1968 and returned to my hometown and a job in the factory. Initially I joined the Army to escape the factory. Worked there till January 1969, quit to sell life insurance with a debit company, Western & Southern Life Insurance, with a route and serviced a specific neighborhood. My clients loved me since I was a Vietnam vet and single. So many eligible daughters that mothers wanted them to marry me and fathers hated me. (I must have reminded them of themselves when they got out of the Army).

Quit that job in September 1970 and my father said if I was not going to work I could not stay in his home so when I came home my clothes were all in the driveway! I rebelled and lived out of my car or stayed with divorced women that had got married and pregnant so the man could avoid the draft. Then the divorces started so I had plenty of women that needed a babysitter.

Started college to draw VA School benefits for spending money, but was not learning anything that could help me to work and have an income. So, I switched benefits from college to enrolling in Cessna Professional Flight School in 1977 and did achieve my Commercial, Instrument, Multiengine Rating. I was then ready to go to work flying the skies. Not easy but did get referred to a new company with one pilot and the owner that started the patrol company as a patrol pilot on AT&T Long Lines. It was great since it was all low-level flying, 300 to 500 feet above ground and limited to only 25 hours a week flying time! Flying the southwest US and bought a sailboat in New Orleans to live on and hoped to work a few years, pay off the boat then sail away.

Next thing I knew New Orleans women did not like tanned men and especially those living on a sailboat, so I invited a girl that I had met thru a friend in Michigan to come and visit for a week and it became a habit. A few visits and I had started begging her to quit her job and live on the boat and share my dream of throwing the lines off the dock and sail the world! That was 1983, then two years later we settled in Naples, FL and have been here since.

I found out I had enough time on the sea to obtain my Commercial Marine License. I then Captained a yacht for charter and worked on private yachts, delivering them to and from the Great Lakes and to the Caribbean Islands. That slowed down till I was just Captaining one sailboat for charters with conventions coming to Naples. I was also driving a horse drawn carriage at night, giving tourists rides thru Olde Naples. I would meet them on the carriage then could also make money taking them sailing and/or flying! Very versatile to survive in Naples.

One night a drunk girl rear ended the horse drawn wagon I was driving. When the car hit us, the horse did not like that and decided he was going back to his trailer and call it a night. There were six of us involved on the wagon and three of us landed in the street! I landed on my hip and fractured my acetabulum, requiring two surgeries. A year later a three-day trial for damages from the girl that rammed us, and the jury awarded me damages for hospital, two surgeries, and a lot of physical therapy.

That ended my captain ventures, but I was able to buy a plane and started flying again with an aircraft sales company. We were going to all the National Business Aircraft Conventions and I was meeting new people and even found a few pilots I had known while at Fort Bragg and patrolling pipelines.

I had bought the condo in Naples in 1992 on a Jack Nicklaus golf course that had many members from Canton, Ohio and Michigan that were seasonal owners. I quit working and decided to start playing golf and enjoying the lifestyle. I decided to get my real-estate salesman license in 2003 and it was the best thing for me as I could finally start making a good living.

In 2005, I found the Cowboys as I had been searching on line for A Co, 82nd Aviation Battalion and found nothing. Then one day, 1975, a girl told me I might find the men I am looking for in 335th Assault Helicopter Co and bingo!

SPANEL, JAMES J.

On 28 February 1972, I left country on a lovely flight home while passing through Anchorage, Alaska with a short refueling stop and a visit to the airport bar. They must have been expecting us since there was a large staff doing the booze service. We consumed a lot of alcohol and stared at the big bear (as I recall it was an upright Polar Bear).

Anyway, we were herded back on the plane for a trip to Travis AFB. The flight was uneventful as it became obvious that altitude and alcohol put us to sleep. At Travis, we all got clean clothes, a steak meal, and were sent on our way home. A few of us took a cab to San Francisco International. As I was heading for my gate, a person not liking my appearance made a rude comment, so I decked him. Shortly thereafter, a couple of MPs took me to my departure gate and babysat me until I got on board.

My next feet on the ground location was in Lincoln, NE where my fiancé met me with her new boyfriend and told me it was over between us. We can all guess how that turned out. Fortunately, I had a close friend attending the University of Nebraska who came to my rescue. I spent a week in an intoxicated condition until he called my parents to rescue him. My parents were not happy with me, but they were glad I was home.

Since my parents lived on a farm in rural Nebraska, I could hoot, holler, and drink all I wanted. My mother became concerned and gave me a dose of her Valium, which of course did not go well with alcohol and ended up in a hospital drinking liquid charcoal. The rest of my decompression was pretty much by myself and isolation from others.

The day finally came whereupon I departed to Fort Hood, TX where I remained until late 1975. My sentence was finally over. While there I met my wife, Jill. She was also in the Army. We got along well and had a quick justice of the peace marriage. We just celebrated out 45th anniversary. We have a daughter who has given us two grand-daughters.

After the Army, I needed something to do. While at Fort Hood I was assigned to an MP unit and law enforcement seemed to get along with me, so I made application and was accepted to the Lincoln Police Department from which I retired in 2005. I guess the excitement kept me somewhat in check, but as my time was coming to an end and would soon be without a structured working environment, it was finally time to visit with a PTSD psychiatrist. I had refused to have anything to do with the VA due to hearing horror stories. After a couple of years with the shrink and a stable dosage of Zoloft, an attorney friend of mine convinced me to give the VA a try. It has gone well so far.

The retirement didn't last long. I had to keep working or else mischief would find me. My current job is with the state department of labor—it keeps me occupied. Structure is necessary for me to stay below the radar. Stupid things still find me, but not as often.

I haven't met up with anyone I served with except for one person. Attending our state Vietnam veterans reunions just don't do it for me so I stay mostly to myself. Keeping up to date with my brothers through the Facebook page sure brings back some memories.

STAPLES, DENNIS LEE

Did you work or go back to school? College after active duty, BS Accounting, CPA, VP-Finance.

What did you go on to do as a career after the war? Continue to fly USAR /Work in Accounting.

Did you get married, divorced, have children? Married /two children/ divorced/ remarried.

Did your military experience influence your thinking about war or about the military in general? NO.

How did your service and experiences affect your life? Positive, enhanced self-confidence. No limits.

Are you a member of a special organization, club, or association? 335th, NRA, Republican Party, Gulf Harbour Country Club.

Is there anything you would like to add that we have not covered? Very happy for the positive treatment our current day service members are receiving. Thank you to all for your service.

After Vietnam, I stayed on active duty. Assigned to Fort Eustis, VA (my choice). In a flying assignment. Continued as a UH-1 IP/SIP. Then Fixed Wing Q C. Headed back to Vietnam to fly fixed wing. BUT war ended, and everyone must get out. My orders changed to fly UH-1 in Korea. It was a good tour. Released from active duty after Korea. I returned to Virginia.

After leaving the service, I returned to college and obtained my degree in business/accounting. Went to work for CPA firm in Norfolk, VA. Designated CPA by the Virginia State Board of Accountancy. After CPA firm worked as a Vice President—Finance for several business operations in the Southeast Virginia area. While in college and while working in accounting, I went into Virginia National Guard followed by USAR as a weekend warrior. More UH-1 IP/SIP time. In 1992, I traded in my suits and ties and accepted a full time IP/SP position as a civilian in the USAR program. This was at Tipton AAF, Fort Meade, MD. Remained there until BRAC closed Tipton and all USAR aircraft moved to Fort Eustis, VA. In 1996, I retired from the military with 30 years of active and reserve service. Designated a Master Army Aviator. Over 6,500 flight hours, almost all in UH-1. During all above resided in Virginia Beach, VA.

After retiring from all formal working positions, I did part time work as a school bus driver in Virginia Beach. They needed drivers and it did not seem too difficult. Driving not a problem, dealing with teenage and younger children is a different story. My second wife Jill and I enjoyed a relaxed world of sailing our 38-foot sloop in the Chesapeake Bay and coastal Atlantic waters. Also, many trips to the Caribbean for bare boat sailing charters. In 2004, relocated to Fernandina Beach, FL. This was something of a business decision for Jill's consulting business. After Jill finally retired from her company, we relocated to Fort Myers, FL. Reside in a mid/high rise condo on the Caloosahatchee River. Travel is the order of the day with occasional trips to visit the children and grandchildren.

Unfortunately, I was unable to attend the 335th Memorial Dedication at Fort Rucker. A trip there is on my list of things to do. I thank all who were involved in this project.

STEIN, JAMES EDWARD

After leaving the Cowboys, I was stationed in Colorado Springs, 43rd Support Group Operations and no flying. "I had too many hours, let others fly." After six months, I went on the G.I. Bill and received a fixed wing instrument commercial pilots license. Also, an invitation with United Airlines Training facility to become an Airline Pilot, "My Dream come true." That would have to wait, I had volunteered early for another tour in Vietnam. I just hoped the War had changed. The last two months with the Cowboys as Cowboy 3, Operations Officer, flying Nighthawk missions, I was investigated twice for unauthorized shootings. Maybe it was just the war in the Delta and the (pacification program).

When I arrived in Vietnam in 1972, I was assigned in the Central Highlands near Dak To, where the Cowboys were in 1967-68 but call sign was Casper. I was assigned to the 7th of 17th Air Cav Ruthless Riders flying the little bird (loach), my call sign "Scalp Hunter Lead". I lasted five months doing what I loved, flying. Our mission was to go out and find the NVA. We would get shot at almost every day. The fifth time getting shot down with the Cav, I was wounded in the knee and after three attempts to save it, they removed the lower half from the knee down of my right leg.

I spent six months recovering at Letterman General Hospital in San Francisco. The Army asked me to remain on Active duty and was sent to my Army Branch career course. But upon completion of the course, I requested retirement and returned to San Francisco.

1975, I was given a medical retirement from the military. For the next three years, I snow skied and golfed all over the western states and Canada. Participating in many local, National and Canadian disabled ski races and golf tournaments. I chaired three National Amputee Golf Tournaments in the northern California summer months. I lived near San Francisco and wintered in Park City, Utah, where I helped the US Ski Team and disabled skiers organize "Celebrate Ski Races" to raise money for the US Ski Team.

1978, living in Marin county near San Francisco, I married Susan. She worked for a charter airline taking long trips allowing me to ski or golf and upon her return if I was on a trip she would join me if liking the resort or mountain. September 1980, we had our first son, Matthew. Susan kept working and I had fun for three and a half years as a stay at home dad raising my son. I had time to study and earned my real estate license, which I used only to buy our first house.

Susan quit working and our friends asked me to join them in their import business. The company imported high end wicker/rattan furniture, baskets, and accessories products.

I started in the front office answering phones. Two months later, they asked (since I had command and operation experience in the military) if I would fill the warehouse operation manager position. They had only a 25,000 square feet warehouse with 20 employees. I took the position.

Creating simple rules to follow, which a few failed to follow, and the warehouse personnel was cleaned up and organized. My second son, Christopher, was born October 1985 and that completed the makeup of our family. As a family, we were able to travel to the islands every year, go camping south of Yosemite and ski Lake Tahoe or Utah. We did all this with two other families with two sons each. Six boys grew up together, about six years apart. Today Matt and Chris are both working and doing well living in San Francisco.

Before I retired from Palecek after 30 years, the company grew to 250,000 square feet, 150 employees, and five showrooms around the country. We also added furniture finishing, upholstery, and lamp man-ufacturing.

I enjoy playing golf and officiating. I've been an NCGA golf of-ficial for 16 years, working Professional and Amateur tournaments in Northern California. Since 2014, I've enjoyed doing tax work for AARP members.

I'm a life member of the Vietnam Helicopter Pilots Assoc., 335th Assault Helicopter Company, 7th/17th Air Cav Ruthless Riders which meet every year, the Novato Elks, Petaluma American Legion, Sons in Retirement, and Santa Rosa MOAA.

STIBBE JR., RUSSEL J. (RUSS)

After I got out of the Army, I moved to St. Louis. I went to work for McDonald Douglas. While there I worked on F-15s. I also got divorced. I managed to work there four years before being laid off.

I then went to work for Emerson Electric as a manufacturing supervisor on TOW Under Armor and Fast Attack Vehicles. They eventually lost the contract and I was again out of a job.

Went to work for Hidden Valley Ski Corp as an outdoor operations manager. Got married to Darlene and we moved to Weston, MO. Went to work for McCormick Distilling Co. in Weston. I worked there for the next 27 years and then retired.

I enjoy hunting, camping, and shooting clay birds. I shoot competition trap for the Fort Leavenworth Trap Team.

Every morning when I get up with my cup of coffee I look at the picture of my B-Model 592. Lots of memories there. The good days and the bad days. Never forget.

My wife and I enjoy helping with the reunions. It's always good to see everyone.

Everybody remembers… on the 8th day God did create the Huey. There is nothing like the smell of JP-4 in the morning.

STILES, HOWARD J.

I took command of the 335th Assault Helicopter Company from Major Paul R. Riley on 25 May 1969. I relinquished command to Vance S. Gammons on 8 November 1969.

When I departed Vietnam, I was stationed at Fort Rucker, Alabama, home of Army Aviation. My initial assignment was with the office of Doctrine, Development, Literature, and Plans. I left the DDL&P office and moved to the Headquarters of the Aviation Center and worked for the Deputy for Personnel (G-1). I left Fort Rucker headed for the Command and General Staff College at Fort Leavenworth, KS in the summer of 1971. I left CGSC in the summer of 1972, headed to Hawaii and the 25th Infantry Division. Upon arrival, I was assigned as the executive officer for the 19th Infantry Battalion. After about a year, I joined the command group for the 25th Infantry Division to be the Secretary to the General Staff. About a year later, it was time for a new job at Fort Shafter. Vietnam was winding down and US Army Pacific was going from a four-star Headquarters, to a two-star Headquarters. I was sent to Fort Shafter to help with the transition as the Secretary to the General Staff.

When getting ready to move back to the mainland, I called Infantry Branch to talk about my future assignment. I said I'd like a ROTC assignment. I was told that type of assignment was not very career enhancing. I replied that I was okay with that, and that I thought I could make a good impression on the students that took ROTC. So, off we went to the University of Wisconsin-Platteville. Platteville was a small town of about ten thousand people and the university had about four thousand students. Our team consisted of three captains and several top NCOs. We developed an outstanding outdoor program that we opened to students that weren't in ROTC. We were well regarded on campus and had a thriving program.

After two years, I was surprised to get orders for battalion command. I even called my boss and said I'd like to stay as I thought we were really making a positive contribution to future young Army officers. His response was, "Howard, don't give up battalion command." So off I went to Fort Bragg to command the 72nd Aviation Battalion. Basically, it was an Air Traffic Control Battalion. At the end of my command tour, I was assigned to the Headquarters, XVIII Airborne Corps as their Aviation Officer. In that job, I met the commander of the First ROTC Region that covered the entire east coast. The various area commanders were colonels stationed at Fort Bragg. The ones that had to go to New England, especially in winter, kept retiring when they discovered the cold New England weather.

One day the ROTC boss (a one-star general) asked me if he could arrange quarters for my family would I consider moving to Fort Devens, outside of Boston, to take command of all New England and Albany, NY ROTC units. Eileen and I jumped at the opportunity as we both had various links to New England. So, again off we went to a new adventure.

I retired from the Army on 31 May 1983. We moved to Maine to a house on Sebago Lake, the second largest lake in Maine. I soon became the broker/owner of Vacationland Real Estate. We were on the lake for thirty years. We thought Maine would be good, but it turned out better than we ever expected.

We took on a variety of volunteer jobs. I volunteered to be on the planning board. Soon I was the chairman. I made the mistake of volunteering my wife to be a room mother and I became the room "father" and built a school playground with lots of help from young mothers and town trade folks. I was asked by the Raymond Selectmen to build a Veterans memorial which was placed along the main high-way through the town. For the Town of Raymond's Bicentennial, my wife and I place 110 flags all over town on lighted telephone poles. The flags were purchased with funds given to the town by local residents. My wonderful Army wife built a pre-school playground for the town, with a lot of help from the residents and by collecting bottles and cans. She started a young mothers group in the village church. One of the most gratifying volunteer jobs I got involved with was becoming a guide/volunteer for Maine handicapped skiing.

At times, we would kid that we either needed to die or move to get out of all those jobs. We did most of our own maintenance on our lake house and eventually said we didn't want to do it anymore. Our search for a retirement home brought us to The Highlands, a senior retire-ment community in Topsham, ME.

We moved in the fall of 2013 and have enjoyed the amenities and new friends at the Highlands. There are many activities to keep one occupied. We have also continued to travel extensively.

STOGNER, DENNIS DEREK

166th Trans Det. 1967-68

After Vietnam, I ended my Army Tour as the Family Housing Supply Sgt. at the Oakland Army Terminal. I had orders for Fort Riley but found a way to get out of that assignment.

I was married while on leave to my best friend's girlfriend. I then went back to Vietnam for a six-month extension.

Marriage produced one child and ended in divorce after 30 months. I then went to school for Jet Turbine Overhaul. I was hoping to work for United Air lines in San Francisco, but there were no jobs available.

I was hired as a fast freight mechanic and was let go while on vacation. The boss hired his son in law to replace me.

Went to work as a mechanic/oiler for an underground construction company. This job ended when the IRS shut the company down.

I then used Cal Vet to join an apprentice welder program, which led me to become a certified welder.

I was hired by FMC Corporation as a welder for MIG Aluminum, Armor, and Stainless Steel. I was given special consideration due to my veteran status and training as an aircraft mechanic. Worked as a team welder on a traveling team with General Electric. Locations included Fort Bliss, Aberdeen Proving Ground, Maryland, Germany, and Belgium.

I switched from Army to USMC support. Worked at Camp Pendleton for seven years as a Field Service Representative.

Returned to the FMC plant and became a FMC Field Service

Representative for Bradley Fighting Vehicle and MLRS (Multiple Launch Rocket System).

I had 25 years with FMC before Ground Systems Division was sold to BAE Systems. Total of 33 years in the same job traveling in support of the US Army M113, Bradley Fighting Vehicle, and MLRS vehicles. Supported the LVTP7A1 Amphibious vehicle for the USMC, ROK Marines, Royal Thai Marines, Brazilian Marina, Spanish Marines, and the Japanese Self Defense Force.

Bought my first Harley Davidson in 2005 and joined the local Harley club. Now on Harley number four, a 2015 Ultra Limited. Retired from BAE Systems in 2009 and have never looked back!

I am a life member of the Harley Owners Group and a life member of the Vietnam Veterans of Diablo Valley.

I finally signed up with the VA after being prodded by friends and fellow veterans for many years. I have now received my disability claim.

We have a Great VA here in Martinez CA.

Married childhood sweetheart in 1978, living happily married ever after!

That's all for now! I am so glad the Cowboys found me!

STOUDT, CHARLES F. (CHUCK)

335TH AHC 12-65/12-66 CWO 2
Falcon, Cowboy, Casper
DEROS – Christmas Day, 1966

Thanks to a generous CO, I got a 19 day drop because the flights (We called them the Blue Ball Express) filled up fast for December 1966. I left Vietnam on Pearl Harbor Day!

My new assignment was Fort Wolters, TX. (anybody know where that is?)

My initial task was as TAC Officer, which was not my first choice, so after a few months, I asked for and was transferred to the Flight Line. I had trained at Wolters in the OH-23D and got a transition at Fort Rucker in the OH-13 while waiting to go to Vietnam. I got my instructor training in the TH-55, (ugh) but was assigned to an OH-13 flight. I was classified as a Triple Threat, IP in all three trainers. There were only five with that classification while I was there. Since Fort Wolters was a critical assignment due to the need for pilots, I did not have to do a second tour in Vietnam. I departed the Army in October 1968, while the war was just getting intense.

I did not have the bad experience's that a lot of troops had on the return to the States, because in 1966 it was still a positive thing. I went to a small town, and from there to the Gulf of Mexico. Not much media there.

1968—I had negotiated a job as an airport manager in a little town just west of Birmingham, AL, so I began my civilian life. My wife was from Tuscaloosa, AL so it was in her back yard.

The business started slow, but I had an 11-year-old Cessna 172, (which I bought while at Fort Wolters) and added a one-year old Cessna 150. Some local pilots allowed me to lease their planes, so in addition to the flight training, I was able to add Air Taxi Operation. I took my FAA 135 check ride in my C-150 and was awarded my VFR Air Taxi Authorization known as "Rebel Air Taxi".

Expanding my business, I negotiated with an AP/IA to set up shop in my hanger. His customer base covered the entire state. I also brought in a couple of fellows to paint airplanes, and the business was "taking off" so to speak. These painters ended up painting all the Factory Lake Amphibian Aircraft for several years. Future freight business was in the works with a BE-18, and a Cessna 310. While flying passengers during a nice busy week, the city fathers could not find me at the airport, (duh) and they put padlocks on my gas pumps, so I cut them off to fuel because I had another flight the next day, (my painters were there to take care of the fuel sales and customers). I got back in the evening, and the electricity had been cut off to my airport. The next day they informed me that my contract was terminated because I was not there to pump gas. No discussion, just get out of Dodge! (Small town, small brains).

1970—I left, flew down to Lafayette, LA, filled out the application, and was hired by PHI, Inc. They were still using a lot of the Icon, Bell 47s, so I was right at home. Loved the J models, as they were the only model in the entire fleet that had a heater. The few JetRangers did not, and it did get cold in the Gulf in the winter.

Upgraded to the JetRanger after 18 months, and had some nice assignments, and some others.

After three years at PHI, Inc. I joined the Air Logistics flight operations. While there, I flew the 205s, (Huey UH-1H) in civilian paint on floats. A sweet flying platform. I also flew the Hughes 500 C.

1975-1978—Becoming bored with day VFR flying in the Gulf, I took a position with an operation that conducted AG flying with helicopters in Indiana.

I worked that operation for the summer, and the next season I leased one of his Ag helicopters, a modified OH-13 with a single seat cabin called the "El Tomcat". Fun little chopper. During my first season, I purchased my own OH-13 G2A1 for my Ag Ops, and since my permanent residence was Mobile, AL, where I wintered, and did pilot services for FBO's there, I put my Bell 47 G2A1 in a flight school there where I got FAA 141 and VA approved for Flight Instruction.

1978—A marriage this summer resulted in a permanent move to Indiana for the Aerial Application business.

The desire to add more airplane operations to my business, I operated out of a little airport in Lafayette, IN, called Halsmer Airport and leased a new Piper Tomahawk for a trainer. After the end of the crop season, I went to Northern Air in Grand Rapids, MI and used my VA bill to get a Learjet Type Rating.

1980—Business was slow, so I went to work at another airport in Indianapolis, IN, and from there joined the National Guard in Shelbyville, IN, flying the OH-58.

1981-1990—While serving in the Guard, I was working for Rhoads Aviation, Inc. located in Columbus, IN. It was a full line Cessna dealer, and a passenger and freight air taxi operation.

I flew just about every twin engine airplane manufactured at that time, including the DC-3. A fun airplane to fly!

1985-Present—The desire to move to warmer climate prevailed, and the summer of 1985 moved back south, to Atlanta, GA. No job, just guts. I took two years to find a flying job, which was at KPDK airport, in Chamblee, GA. They needed a Beech 18 pilot, so I sort of mislead the Chief Pilot about my experience in the type and got the job. They had two of them, and they were both tri-gear models. Who can't fly a Tri Gear airplane?

My actual time in the Beech-18 was "0". I later worked for Smith Air, which operated another tri-gear Beech-18, as well as conventional gear, and beginning operations in a Lear 23.

I accumulated over 1,000 hours in the Beech-18s and another 1,000+ in Learjet's over the next three years.

Taking a break from flying for a couple of years, I was an assistant manager for The Great American Cookie Company store at South-lake Mall, and I was a manager of the largest volume Starvin' Marvin Fuel Center, just south of Atlanta.

1992—The desire to get back in the cockpit resulted in my employment with Aviation Atlanta, Inc., the largest flight school in the Atlanta area located at KPDK, as Chief Flight Instructor.

A brief stint as a free-lance pilot in the area, I flew the daughter and family, of the family that started the Varsity at Georgia Tech, and began flying for an operation that moved cancelled checks and mail from banks in the Southeast to the Federal Reserve in Atlanta. Flying out of KATL in a Baron-58 daily, departing at 0700 and arriving about 1700 five days a week, then on Sunday a trip to KMEM and back.

We started operating a KA-200 for passengers, and things got real busy.

Flying all over the Southeast, the Midwest, and several trips to the Bahamas kept me quite busy. Add in the week-end trips with gamblers to the casino's in Tunica, MS and Biloxi, MS.

2004—After a tragic accident closed the doors of that operation, I went to work for one of the former customers that bought his own airplane.

It was a KA-350, the largest general aviation turboprop airplane, and I was operating it in a single pilot mode. It was a family plane, so mostly family trips, like to Panama City, FL for week-ends, a trip to Costa Rica for a week, a trip to the Virgin Islands, and a trip to Alaska for a week, (the day I landed in Anchorage, AK they recorded a record high temperature, and it was the second week in September).

2005 — Another opportunity flying a KA-350 developed closer to my home, (I had been driving 75 miles each way to work) so taking that position, another pilot was also hired so we would have a good cockpit. Turned out to be a retired Army Aviator. Vietnam in 1968 (he went through Wolters while I was an Instructor there), again in 1971, that time in the King Air. He flew those for the duration of his hitch, retired after 20, went to work for United Airlines, retired from them at 58 yrs. old, and went to work flying with me in the KA-350.

We had a really good job there as the boss used it for his gambling and hunting trips. I think we did three or four trips for business in four years. His friend would use the airplane on most week-ends, to go to Daytona Beach. We would go play golf, Hooters Wings, and Beer. Tough Duty!

2009 — The other pilot got schooled in the Beech Premier 1A Jet, and then the Cessna CJ-4. I would go on the trips to be a safety pilot in the event of . . .? We had almost 100 years, and 55,000 hours of safe flying in the cockpit.

2016 — My last revenue flight was in August of 2016, so October 1965 to August 2016 was my commercial flying career of 51 years and right at 30,000 flight hours.

I am soon to be awarded the FAA Master Pilot Award, which is an award for 50 years of flying with no accidents, and no violations.

Enough about me. I am married, second wife, have four grown children, and four grandchildren. My oldest son, born while I was in Vietnam, served 23 years in the Air Force, starting out as crewchief and ending as top Sergeant of the 94th Fighter Squadron. One of the elite of the Air Force. He has one son going to Virginia Tech.

Son number two was born while I was stationed at Fort Wolters, TX.

He went to a civilian school for Aviation Maintenance and is the top mechanic at The Maintenance Group at KPDK, a very big jet maintenance facility. Two children, daughter in college and a son in high school.

Daughter one is a school teacher in Stuart, FL, has a daughter, now graduated from college, and loves her work.

Daughter two graduated from California College of Art and is pursuing her chosen field.

Presently I am a member of:

Distinguished Flying Cross Society, QUIET BIRDMAN, American Legion Post 127, Civil Air Patrol KPDK Senior SQ 130, VHPA Life Member, GA-VHPA, COMBAT HELICOPTER PILOTS ASSN, AOPA, EAA National, EAA KLZU 690, Georgia Commemorative Air Force, Silver Wings Fraternity, and NBA.

I have never regretted my service to the country as an aviator and have great respect for all those who choose to serve our country in any military capacity. I received some valuable training, and experience from my tour, and was able to utilize that experience in my varied aviation activities.

STRIBLING, NEIL PORTER

After Vietnam, I continued to serve until retirement in 1978. Entered the Army as a private and retired as a major. I then taught JROTC in the SC School System until retirement in 1998. I have two children, Frank and Mona. Frank retired from the Army as a 1ST SGT and now works at Fort Jackson, SC. Mona is a counselor at the University of Southern Florida in Tampa.

SUPPOK, REGIS JOHN

After Vietnam, I completed my obligated service at Fort Meade, MD in the Air Cav Troop of the 6th Armored Cavalry Regiment. I was training officer, then operations officer in that until separation.

After leaving the service, I moved to Alaska and worked for ERA Helicopters as a pilot, serving the oil industry and other operations like firefighting and logistical support. This was interesting and satisfying but involved prolonged periods of separation from my family (three children). I found myself wishing more home time, so began searching for new opportunities that did not involve separations. This resulted in finding an opening as a high school teacher in the Anchorage school system teaching math and science and helping coach a high school rifle team. After two years, I learned that the Alaska Army National Guard was expanding to include an Assault Helicopter Company. I applied for membership in the Guard and was accepted. I was the senior officer interested so was appointed to be the unit commander of the 1898th Aviation Company (Assault Helicopter) with the mission to organize, recruit members, establish training, and conduct unit drills. The unit grew from six officers and men at the first drill meeting to full strength with 31 UH-1 slicks and guns in two and a half years. We began receiving UH-1s about five months after organizing. It took more than two years to get the full complement of aircraft, but getting them increased our visibility and certainly helped get new members, mostly veterans who wanted to continue serving and flying. I left teaching to become a Technician (civil service) in full time employment with the Alaska National Guard. After about two and a half years, I was reassigned to State HQ as State Aviation Safety Officer but continued fulltime flying as Technician Flight Instructor, then Aircraft Maintenance Supervisor, and finally as State Army Aviation Officer. I qualified in fixed wing aircraft assigned to Alaska in the first years of my service in the Guard and flew U8D, U8F, U21, C12D, C12F, U1, U10, and UV18(twin otter) all over the state. What a wonderful experience that was, seeing the wonders of Alaska, working with villagers, and dealing with weather, arctic cold, and the remoteness of some installations we dealt with.

I really enjoyed living in Alaska with its varying conditions, outdoor opportunities for fishing, hunting, sightseeing, camping, and exploring. Often good things and lifestyles come to turning point and mine did in 1987. Family obligations and changes dictated moving back to the lower 48, so I found new employment as a Department of Army civilian pilot at Rock Island Arsenal in Illinois with a later transfer to pilot at Letterkenny Army Depot in Pennsylvania where I retired from civil service when the pilot positions were transferred to other organizations.

In Illinois, I served in the Army Reserve at a USAR School, teaching in the CASS3 program. Upon moving to Pennsylvania, I retired from active USAR service because I could not find a reserve position in the area. I am now completely retired as LTC. My love of flying and teaching led me to my final employment at the University of Illinois Institute of Aviation as a full-time flight instructor until that program was terminated in 2014 and I again retired. I think I had truly found my niche there. I really enjoyed flight training with college students, interacting with them, and renewing the wonder of seeing them experience flying, grow in skills, and helping them progress in their chosen field. I miss that interaction now but have compensated in other ways.

After retiring, I am travelling with my wife of 53 years, Janet, in our RV, playing music (mountain dulcimer) at local nursing homes and assisted living facilities, fishing, hunting, and just generally enjoying life as it comes. Seventeen years ago, I was in a motorcycle accident that resulted in my left leg being amputated above the knee and my right leg being fused at the knee. This limits my ability to walk somewhat (no long distances and unable to climb ladders: steps are OK) but I was able to continue flying single engine airplanes with hand controls for rudder, no multi-engine, and no helicopter. I have been active in working with other amputees, helping to inspire them to accepting new limitations, learning to work with prosthetics, and discard the despair some feel at their changed life. Fortunately, I have never experienced despair, remorse, or feelings of inadequacy over my situation. I view it as a challenge God has thrown at me to deal with and I try to remember that He also gives me ways to meet that challenge. It mostly involves thinking through ways to do it (military problem solving).

TEASDALE, THOMAS C.

After Vietnam, I remained in the Army and returned to Vietnam for the second time in 1971. I also went to Iran in 1976 for a few months. Finally retired at Fort Rucker in January 1979.

Before retiring, I started to attend church and chose to serve the Lord as a pastor, so my wife and I started a church in Enterprise, AL, while still in the service. In December 1979, we packed our bags and moved to a small Alaska native village and have been here for 34 years. It has been exciting for us. My coldest day here was 68 below zero. The winters are long, the summers short. We did take a few years break to care for my wife's folks until they passed away, then we moved back to Northway.

Throughout our 60 years of marriage, we've had nine children and fostered almost 200 hundred. With our children, we have 17 grandchildren and eight great grandchildren. My wife had our fifth child shortly after my arrival in Vietnam.

My military experience did not influence my thinking about the military since I wanted to be in the Army since I was six years old.

Due to our way of life here in Northway, my military life has played a small role in things I've done. Northway was a military base during World War II, with 5,000 soldiers stationed here. At times the people here find old military items such as smoke bombs, ammo, old military vehicle parts, and my experience has helped to identify items and reduce any fear that the unknown items may cause.

I'm a member of USAA and a Church of God organization currently. Being almost eighty years old, retirement has entered my mind.

An interesting family item: Last year in December 2016, I found out that I have a sister that I never knew I had. She lives in Montana. Several years before that, I also found out I had a brother, also in Montana, I didn't know about, but did meet him. He passed away from wounds received in Vietnam.

TROTOGOTT, JOHN A.

I got out of the military at Fort Lewis, WA. I stayed with an uncle (Vancouver, WA) for a few weeks elk hunting on horseback, fishing, and camping on Mt St Helen's.

I then went back to Warren, OH and attended Youngstown State University while working at Republic Steel as a millwright. Got married and divorced within three years. I owned a bar while going to school and still worked in the steel mill. However, I got itchy for the Northwest.

Nine years later, I left Warren, OH in 1978 with a buddy who was going to San Francisco. I stayed with his family for a while then headed north to the Portland/Vancouver area. Then back in a steel mill for one year. I ran a flooring warehouse for a neighbor while in Portland.

Got married to my Ohio sweetie in Vancouver, WA. Then took a job offer at a titanium plant in Henderson, NV. While in NV, I received a degree in the arts (still a millwright). I initiated and designed an industrial style curriculum for a college. After 12 years there, I just had to go back to the Northwest.

Ended up in Roseburg, OR selling industrial power transmissions items to the lumber industry, etc. Before retiring at the age of 62, with both knees replaced, I was selling personal safety equipment.

As for Vietnam, we (the USA) all got sucked into this civil war. Many lies, as we all know now. A few of us were aware of this while in Vietnam. However, we had a duty, we got called, and we went to defend our country. I will never forget my time in Vietnam and all the

231

great people that I worked with. They truly are my friends for life and I love them all.

I miss the ones that did not come home. I always prayed that the USA would not get involved again but as we all know. . . that it will never stop.

UNGERER, FRED RICHARD

After Vietnam, I was stationed at Fort Hunter AAF in Savannah, Georgia for the remainder of my military service (September 1968-January 1970). While there, I was sent to IMOI (instrument Instructor) school and after that, did IMOI training for the remaining time. I was scheduled to return to Vietnam as a Cobra pilot in April 1970 when early outs were offered. I took the early out and departed at end of January 1970.

After leaving the service, I looked for work as a pilot, not finding any in the US, I ended up working mostly overseas for the better part of ten years for most of the major corporations that were in existence at that time. Spent time in South America, Indonesia, Saudi Arabia, and Mexico doing various contract work, mostly for the oil, construction industry, and a helicopter manufacturer. These jobs consisted in large part, of doing off shore, Long Line, Heli-rigging, oil survey, construction, military maintenance, and training in desert and jungle conditions.

In 1978, I returned to the US permanently when I acquired a job as an EMS pilot for a county in Florida. Married my wife Gail in 1980 and a new life started again. The contract was just starting out, and over the course of twenty-five years we built it up from use of a Bell 206 JetRanger to a MBB 105 and, eventually acquiring an EC 145 as a single pilot IFR certified 24-hour operation. I retired as chief pilot in 2003.

During that same period, I also worked part time as a mosquito control pilot operating DC-3s and Bell 206s. During this period, I acquired my type ratings in both the DC3 and Learjet. Bought a few aircraft over the years to enjoy as well as doing some part time aerial photography. The only way I could afford to own one.

After retiring, I continued to do some part time mosquito control work for a few years than decided to hit the road in my RV. My wife Gail, enjoyed the life style. When she passed away in 2006, I continued doing it each summer since. Have enjoyed many new places in the process. Now I spend the winters in Florida in a home I built in 1988 in an airpark (another dream I always had), enjoying building my airplane, a Glasair RG, to which I hope to have flying sometime this year. It has only been a 33-year project. Life just kept getting in the way, you know how that goes, I'm sure. Figured it was about time to start emptying the bucket list, so I acquired my seaplane and glider ratings. Did not think I would get hooked on the gliders, but sure did. Proved to be an extremely satisfying form of flying, so I bought into one and been enjoying that as well. Have flown over twenty types in over a dozen different locations. Now that the twilight years are approaching, I can honestly look back and say that I have had a full life in an avocation that I have enjoyed since childhood, with a partner that enjoyed it as much as I did. Can't ask for more.

UTZMAN, CHARLES D.

Aviation Officer, 173rd Airborne Brigade, January-June 1967
Commander, 335th Assault Helicopter Company June-November 1967

I'm using a very old faded black print on faded green paper incomplete copy of my DA Form 66 as a reference for my history that follows. I'm also using memories from my 88-year-old brain which, for the most part, is in pale gray, or red, blue, yellow whichever, pale is the key word! Hopefully what I end up with on paper will be reasonably correct. But I must remember the many times I've said over the years that this old veteran doesn't tell lies about what I've done, but because I sometimes forget details I just might "embellish" my stories a little bit to keep them interesting. You can take that as a warning!

I joined the 173rd Airborne Brigade in Bien Hoa on 16 January 1967 as Aviation Officer, BG John R Deane, Commanding. Working for and serving with Gen Deane (Uncle Jack) was a very real experience in leadership. He let me do my job. He expected me to do my job. And when I erred, he was quick to critique me and just as quick to suggest how I could have done it better. His "grunts" were his primary concern, no matter the circumstances. He loved, respected, and took care of his grunts. I took great pride in being one of his troops. He visited them during operations in many areas where we, his staff, and Caspers, didn't want him to go. I treasure the Combat Infantry Badge and bronze star on my senior parachutist badge I "earned" with the Herd. Makes me feel like I just might have come close to qualifying as a real grunt. Gen Deane retired with four stars—I stayed in contact with him until he died (cancer) a few years ago.

While I was a very new and nervous FNG, I accompanied Uncle Jack on one of his "check on his grunts missions." We received a message from one the units that they had captured a POW! He told them he would pick him up. We did and as he sat on the floor of the helicopter in front of me—we locked eyes! My thought: this is our enemy, my enemy! A handsome man, dark brown eyes, dressed in clean and fitting black and white clothing. He looked harmless! The thinking of a FNG. I found out soon that he was indeed our enemy, my enemy! I can still picture him.

On 10 June 1967, not long after the mid-air collision of two Cowboy slicks en route on a mission, I assumed command of the Cowboys on a very muddy corner of Camp Enari during a rain shower. Why were we given this somewhat bare and isolated part of Enari? You think it might have had something to do with the Cowboys having announced their arrival with the ill-timed, ill-aimed, and unplanned launch of a rocket on short final! No wonder the 4th Division people seemed to not appreciate having us in their midst!

We moved to Dak To to co-locate with the 173rd. In sandbagged tents along the runway! Some discomforts, obviously, but also a relief to no longer be at Camp Enari! I don't have a fitting word to describe the relationship we had with the Herd. We took care of them—they took care of us! Under any and all conditions. (Even to giving us a quota for participation in the activities in Dak Kok City at the end of the runway! Did Dak Kok City really exist? Yes, it did. (I have an unofficial document entitled: "Regulations and Penalties for Dak Kok City—Republic of Vietnam" Minimum for visitation: NV$500.)

The Cowboys! Good people! Good troops! Good attitudes! Dedication! Loyal! Determined! Caring! Reliable! Concerned! Proud! Professional! Successful! A ***TEAM!*** (in bold, italicized, and underlined all-cap letters for emphasis)! Teams within a TEAM! "AWESOME" is, in my opinion, an overused word but I'm using it to describe the Cowboys! AWESOME! Absolutely! But even "awesome" is inadequate.

LRRP, combat assaults, medical evacuation, search and rescue, hot meals, resupply, recovery of casualties, night missions! Short missions! 24/7 missions! Mail and coke missions! Beer missions? Some of those, too!

So many memorable events/missions/operations! Like the combat assault in the Dak To area that involved a B-52 strike, followed by an airstrike to establish an LZ. Both failed! So, we rappelled grunts into a bomb crater to enlarge it for an LZ. A grunt got hung up on his rope about halfway down—couldn't go up, couldn't go down! Returned him to base camp dangling from the slick at 1,200 feet! Followed by me on the left side and General Deane on the right—until we got him to the pickup area where he was gently lowered to the ground. Mission accomplished!

Resumed combat operation! Routine? No mission was routine—each was unique! I never knew the identity of that young grunt—he had quite a story to tell his grandkids—I've wondered about his future. Just as I wonder about the future of the young grunt that got hung up for hours in the huge tree in the middle of the drop zone of the 173rd combat jump on 22 February 1967!

Heroes—ALL! Cowboys, Mustangs, Ramrods, Falcons, Horsethief.

I left Vietnam with the conviction that my tour with the 173rd and Cowboys was the most special and professionally satisfying assignment of my career! In fact, my life! And no future assignments could possibly be better!

Next assignment: CO, 3rd Battalion, Troop Brigade, USAPHC, Fort Wolters, TX. A very puzzling and challenging chore. Mix a bunch of permanent party people, students, civilians, contractors, etc., and what did I get? A brand new and different situation from anything in my past. Turned out to be fun in strange kinds of ways. A special perk was in being reunited with a special buddy from earlier days at Fort Rucker—Vance Gammons.

Fort Wolters didn't last long. 15 May 1968 I was "specially selected, no refusal" to join CDSO, War Games Section, Studies Branch, Study Directorate 3, HQUSACDCISS, Fort Belvoir, VA. A study group to develop the concept, production, and employment of the ultimate US Army Armed Helicopter. Interesting? Challenging? Not really. Okay, a little. Two perks: getting rated in whatever our twin-engine fixed wing airplane was at the time. Didn't need the rating. Didn't want it either. The second: a beautiful A-model of the ultimate gunship—by Lockheed.

I volunteered to return to Vietnam July 1969—Deputy Aviation Officer, HQIFFORCEV, USARPAC, Nha Trang. Another staff job but a good assignment at a higher staff level. It was interesting to experience being a part of the planning and use of various resources at field force level.

On 21 January 1970, I became Dragon 6, commander of the 52nd Aviation Battalion and commander of Camp Holloway, Pleiku. After being at Nha Trang for six months, I thought I was prepared for the 52nd and Holloway. I was wrong. Very wrong. I was just somewhat prepared. Downright scary. Chinooks, Hueys, gunships, 2,000 plus troops, civilians, multi-locations, security, maintenance, airfield operations, combat support operations, perimeter security, bunker security and maintenance, administration, enlisted and officer clubs, banana tree, post exchange…And more! Much more! To include our very own area of operations—secured by what amounted to our own light infantry company. And two tigers!

On a reasonably quiet morning at my desk—I got a warning about something going on in the hooch area. Discovered a very large, beautiful, but very dead tiger strung up between two hooches! Surrounded by numerous onlookers—all of them excited and noisy. Including a bunch of Vietnamese hooch maids. Seems a crew returning from a single ship mission spotted a tiger at just about the same time they needed to test fire their guns. Unfortunately for the tiger, it was in the line of fire. So, they brought it home. Of course, I had to do my thing. Took the AC and his crew around a corner and made some remarks about compromising safety of crew, helicopter, mission, and a few other things. Dismissed them and returned to my desk. Upset? No! Huge morale booster! They had just joined my growing number of young heroes as Super Heroes. I never mentioned it again.

The second tiger? Another large, beautiful, and very dead tiger, but this one stuffed—compliments of Sergeant Major Roman Chomskis when I left the battalion. No, not the same tiger. Story goes that he bought it from a Korean vendor in the area who wanted a contract to operate on Holloway. I got the tiger, but he didn't get a contract. Quartermaster people shipped it home for me, I still have it—named it Charlene. It's beat and battered, like me now, but a survivor. From being in displays, parades, libraries, cheering sessions, booster club, football games, giving bareback rides.

We regularly supported the 4th Infantry Division, ARVN, Special Forces, and all others who might be operating in or passing through our area of operations. Our 57th AHC at Kontum almost exclusively supported special operations within and beyond undefined boundaries.

One of my favorite chores—visiting the airfield perimeter bunkers. At night! An opportunity to tell some of our special young soldiers how much I appreciated them for their performance of a very tough, disliked, and all too often unappreciated duty. Theirs was tough duty. Tough duties during the day at their "regular" jobs followed by night shifts in the bunkers and numerous other tasks. Crewchiefs and door gunners—up very early to see that their ships were ready for seemingly always early morning missions! Flying all day! Then putting their birds to bed at night!

Not many dull or idle moments.

We had an adventure with an air cavalry squadron that moved onto Holloway for a few weeks. Not a problem dealing with logistical and security support for them. But a problem dealing with the wearing of hats and behavior in our officers' club. And a dog! Our guys resented the hats, the behavior, and the dog in the club! The dog! One evening during a visit at the club, I, on impulse, physically and emphatically removed the dog through the swinging front door. Total quietness! Message delivered! Never saw the dog again! Had to take care of my troops! Not funny at the time, but became laughable in the long run! Never did completely resolve the big, black hat problem. Not much of a heroic moment for me but I did get some good feedback.

After Pleiku, I was assigned as aviation officer, 3rd Armored Division, Fort Hood TX. A good and enjoyable assignment but not a very fast paced or demanding one, especially compared to the one I had just left. But that changed. I became commander of the 1/50th Mechanized Infantry Battalion, 2nd Brigade, 3rd Armored Division, Fort Hood. Young company commanders, young officers, young troops—interspersed with scattered older soldiers. Even a few old soldiers! Peacetime attitudes! Peacetime resources! Peacetime maintenance! Peacetime problems! It was good duty—I enjoyed it—with much appreciation of the performance and dedication of our young soldiers. But DR's (delinquency reports), traffic accidents, DUI's, etc., spoiled some good times. And the personnel records of some very good troops. Too many instances, I thought, where young troops were judged and punished too severely for youthful actions that could have been classified as normal.

I had been a not so carefree bachelor, enjoying a life without the complications and responsibilities of marriage and family. Then on July 5, 1970, I got a wife, two kids, a dog, and a house mortgage because of two little words: "I Do!" Then on January 16, 1971, I got another blessing—a daughter! My wife had been an Air Force wife—her husband a navigator on B'52s then as RSO on the SR-71 Blackbird. The most exciting aircraft ever? And "extraordinary" missions! He had flown very high and very fast. Quite an adjustment for her in dealing with me who had flown so low and slow! (He was killed in a traffic accident at Beale AFB—a pickup pulled in front of him on his motorcycle!)

After Fort Hood, we took our family to Germany—Heidelberg! Another not very exciting high-level staff job. We had many great adventures in Germany and Europe. After a year, we opted to retire after 25 years' service and returned to Austin, Texas. Our primary interest and activity—getting our kids through high school and onward to lives as they wanted to live them. I had planned on being a high school counselor, but found too many well qualified people standing in line for jobs. Instead I got an elementary school teaching certificate. Then had ten years of service as a fifth-grade science teacher, a second and third grade gifted and talented teacher, and computer teacher for grades three through five. Ten great years! Loved the kids, loved teaching, loved the schools, appreciated the parents. A hard-nose former grunt, paratrooper, special forces, helicopter junkie teaching elementary school! Yes! Then politics got in my way and I reluctantly left teaching.

After completing her program at Texas Tech, our daughter settled in the DFW area. We followed her. Then she got us a Texas Aggie son-in-law! And two grandsons, now nine and eleven. We're very involved in their lives and love and appreciate every minute of it!

We live in an "active adult" community near Argyle, Texas. My wife is involved in many of the women's activities. I mess around and play pickleball! I judge myself physically fit and enjoy being competitive with my juniors on the pickleball court. Life is good! God is good! Many blessings!

Pre-Vietnam, 173rd and Cowboys.

I grew up very country in central Texas—enlisted Army infantry in August 1946 at 16 years old. Basic at Camp McClellan, AL. Then got lucky and won a two-week cruise to Yokohama on an old World War II worn out troop ship, along with about 3,999 others! Two weeks of poker, crap shooting, and puke! All over! Everywhere!

I signed on for the 11th Airborne Division, influenced a little bit by the promise of a 200-foot enlisted club bar! and beer! No, not correct—I signed on after being influenced by buddies I had accumulated along the way—following their lead because I didn't know any better! Then I discovered that I had signed on to jump out of airplanes! Old and worn out C-46s and just as worn out Army pilots! Didn't know any better! But after the third jump, I knew better! Scared beyond "s———less"! By then I knew well what I had gotten myself into, but the consequences of failure were worse than my fear, so managed it just enough to do the required five jumps to qualify. Fun? Depends on how you define "fun!" I had become a badass paratrooper and could now drink beer at the two-hundred-foot bar! My duty: company clerk of a parachute infantry company, 188th Parachute Infantry Regiment, 11th Airborne Division, Sendai, Japan! Good job, knew everybody, knew everything going on, worked for the First Sergeant and Company Commander, got lots of time in the dayroom playing ping-pong and because I was very good doing the personnel morning report, they wouldn't fire me! At 17-years old—a good life. Until three months later when I had to make a sixth jump to qualify for jump pay! Came home a T-5!

Life was not good for the Japanese people. I took great pride in our having won the war with Japan. I did not take great pride in what we had done to the Japanese people. Devastation! Hunger! Too young to fully grasp the evidence that war is a terrible solution to differences! Discharged at the end of 1947. Another luxurious cruise back to San Francisco...this trip almost fun. Standing at the bow of the ship at dawn, watching the Golden Gate Bridge become visible as we approached was beautiful! Home!

1950 Korea happened. I reenlisted for service in Korea—got

stopped at Fort Ord, California. Company Clerk of a basic training company, 8th Infantry Division…Fun! But not the Army infantry stuff I really wanted. Applied for transfer back to airborne duty. Got It! 101st Airborne Division, Fort Campbell, KY, Division G3 section. Clerk!

1951—applied for and got Infantry OCS, Fort Benning. Second lieutenant, Infantry. Proud accomplishment! Found myself a heavy weapons platoon leader, 504 Airborne Regiment, 82d Airborne Division, Fort Campbell. Fun and games! Already knew what NCO's did so relied on some very good ones to break me in!

Going through Tokyo in 1953, I signed on for a unit so secret nobody could talk about until after attendance at a classified briefing. Took an assignment as an advisor to a partisan battalion of north Koreans with the 8240th Army Service Unit, FED/LD(K), duty station on islands far west of Inchon with operations into north Korea. Fascinating, challenging, and fun. Then, on short notice, we had to deliver our partisans, without their families, to the south Korean Army. Seems, even under our control, they were too much of a threat to the south Korean regime. Ugly stuff! Sneaky stuff!

To continue a special forces capability, the United Nations Partisan Infantry Korea was born. Korean soldiers from the South Korean Army—volunteers for special duty. Wolmi Do island just west of Inchon. Jump school, basic training, special forces training. More fun and games—especially our jump school. From our island, I sailed into Inchon harbor and tidal basin in boats ranging from an Army assault boat with a 25 HP motor to LST's.

I returned to US with assignment to 77th Special Forces Group, Fort Bragg. Fun and demanding training operations in North Carolina, Louisiana, Pisgah National Forest, Camp Lejeune, and Colorado with people, many of them Europeans, with amazing training, backgrounds, and experience. Good times being sneaky in SF when we first got our Green Berets. With them, too, when we had them taken away from us because we "misbehaved" during an 82d training exercise. CG/82d fired us! Kicked us out! Successfully "re-earned" them, but it took a while.

1956—57—Flight School at Gary AFB TX and Fort Rucker! Two years? Slow to solo! No, a broken arm in an L-19 accident. Duty with the 101st Airborne Division, Fort Campbell, KY, followed. Routine for the most part—good practice operating off short strips, roads, flare drops, etc., routine until on a Sunday afternoon training flight I misjudged, causing a quick decision whether I fly under a wire across the road or fly over it. Maybe! Flew under—lesson learned! A second lesson when I inadvertently, clumsily, and fumblefingeredly dropped my flares from a very low altitude. In fact, on the ground, during run-up, at the end of the airstrip. The flares were effective though—lit up the whole division headquarters. Another lesson learned! Broke an H-23 helicopter at Wolters during RW qualification! Broke it bad! Totaled! IP demonstrating a pinnacle takeoff on a hot summer day! Sudden stoppage but a soft landing—in a stream bed with about 18 inches of water! No injuries! Except for pride!

1959—Advanced Infantry Course, Fort Benning. R&R!

1960—1962. MOI, flight instructor, flight commander—H-34s. Project Officer, Department of Tactics, Helicopter gunnery. Developing hardware, armament, training manuals, personnel requirements, firing range procedures. More fun and games, mainly because of the people I worked with and as an SS-10 and SS-11 wire guided missile instructor. Adventures in life with Vance Gammons as RW instructors and in the helicopter gunnery business.

1963-1964 A ground assignment followed with the 8th Infantry Division, Baumholder, Germany. Infantry company commander, 16th Infantry Battalion, S2 and S3, 2d Brigade, 8th Infantry Division. Good assignments, good experience. Worked for a great brigade commander: Col. John G Wheelock III—tough but good tough. I learned a lot from him. But an instance of hovering an H-13 for hours for flight pay on a foggy day was dumb and dumber! Didn't happen twice! My tour in Europe ended with an assignment with the Aviation Section, HQ 7th Army, Europe, Stuttgart.

1965 Command and General Staff School, Fort Leavenworth, KS. St Benedicts College, Atchison KS, to complete a degree. Discovered that I could still learn! Graduated "summa cum laude!" but since we were their first experience with bootstrappers, and they liked us, it was more like "gratis cum laude!" Rewarding experiences! After completion — en route Vietnam.

1967! Did my pre-1967 experience prepare me for the 173rd and the Cowboys? Yes? Partially! Especially in dealing with troops in various environments and circumstances. No? Partially! How could any military and/or personal experiences fully prepare anybody to serve with and for the 173rd and Cowboy Team! A complete team at all levels and positions! Professionally competent! Professionally led! Dedicated to mission accomplishment! Dedicated to units and comrades! Utmost loyalty to units and comrades! Comrades! Comrades at all levels! The young people of all branches and services are, have been, and will always be, my heroes! Pilots, crewchiefs, mechanics, ammo bearers, clerks, et al! And the grunts! Especially the grunts! Our beloved grunts! Respected grunts! Trusted grunts! Devoted grunts! Sacrificing grunts! The toughest of all jobs! ALL jobs! Dedicated to mission accomplishment! And to each other! Dedicated to taking care of their comrades to the "nth" degree no matter the conditions and circumstances! I extend my most sincere thanks and appreciation to all grunts for your service! It was my great honor and privilege to have served with you and for you!

Referring to my comments about my conviction that no other assignment could measure up to my tour with the 173rd and the Cowboys…My conviction has never changed, perhaps a slight, a very slight, waver now and then, but never changed!

And all my comments about the 173rd and Cowboys? ***BOLD, ALL CAPS, BLACK ON WHITE AND ITALICIZED! (for emphasis!)***

VANDERPOL, JERALD W.

After Vietnam, I was sent to Germany for my last six months of service.

Went to work at Hamilton Standard (Division of United Aircraft Corporation) as a Quality Assurance Inspector. They had contracts overhauling UH1 transmissions, 42°, 90° gearboxes, T53-Lycoming Allison 250 engines and repairing main rotor and tail rotor blades.

Married December 7, 1973 (still married forty-four years later).

Started to go to college at night on the G.I. Bill. Long road working full time and school, but finally graduated.

Continued to work for Hamilton Standard and was promoted to various positions within the company as Manager of Propeller and Rotor Department; Fuel Control Hydro-Mechanical Department Manager; and finally, Operations Manager of our facility in Long Beach, California.

During this time, I was also in the Reserves with the 336th Assault Helicopter Company as a Crewchief on an "M" model gunship and was also in charge of armament for our gunships. Every other drill we would take our gunships to Camp Irwin to work out on the impact range. Loved all the stick time I got as we were always short on pilots.

During this time, we had two boys born in 1970 and 1981. In 1982, I decided I could no longer live in Southern California. Wanted a better life for my boys to grow up in! Moved to Northern California into the Sierra Foothills. Bought a home on ten acres where my family could grow up in the great outdoors. Hiking, fishing, camping, skiing.

Went to work for a small company called Tri Tool, Inc. as Manager of Engineering, bringing aerospace technology with me! After three years became CEO of company!

Traveled the world on business over the next twenty-eight years working on projects for DoD, DOE, Navy Nuclear, solving problems by designing special tools for remote high radiation applications. We even designed and built special tools for the international space station, B1, B2, A10, C-17, and V-22.

Volunteer fireman, engineer, and EMT with local fire department.

Accomplishments: Eighteen patents on products with the United States Patent Office.

Retired July 2013. Enjoying life traveling with my family.

WAKEFIELD, MICHAEL DENNIS

Career: After Vietnam, I spent a great deal of time trying to figure out what educationally, I wanted to do with my life… not an easy task. I went to Italy for Medical School at Perugia University, quit, came home to the USA; and then graduated from University of La Verne with a BS in Biology. I then went to and graduated from University of the Pacific with a Doctorate of Pharmacy degree (Pharm. D). I attended Mercer University for post-doc Nuclear Pharmacy Training and have had phenomenal success in my professional life. I am a board certified nuclear pharmacist (BCNP), one of only 500 in the USA, and I worked a long time as a Nuclear Scientist for an Oncology Company, Biogenic-Idec. I managed numerous nuclear pharmacy facilities in California, and now enjoy just being a regular pharmacist.

In 2011, I moved from San Diego, CA to Fort Campbell, KY where I worked four years as a civilian pharmacist for the US Army, a very rewarding and moving experience for me… back in the Army… go figure. I transferred back to CA, I now work for the US Navy as a civilian Navy DoD pharmacist at Camp Pendleton and enjoy supporting the USMC and US Navy… I believe in one way or another, I will always be supporting the DoD and our military… proud to be part of military affiliations and a very proud American.

Family: I married late in my life and divorced 17 years later. In 2013, I married again, the love of my life, and I now live in Oceanside, CA, close to USMC Camp Pendleton.

I have four children, now all grown-up, from my first marriage,

Laura, Mary, Sarah, and Warren. I have one granddaughter, and another one on the way.

In 1994 (at 44 years of age), I suffered a heart attack and survived. In 2012, I suffered more cardiac related events and wound up getting a pacemaker. I am grateful for the VA and my VA disability and thankful for their help… I am a survivor many, many times over.

WAX, GORDON WILLIAM

I grew up on my parent's Montana farm where I learned how to work hard, be honest, and do most anything. But, boy! How the times have changed from our farm house being the only one on the road to now over seventeen houses. In June of 1967, I joined the Army and graduated from the Huey school at Fort Eustis. After receiving orders for Vietnam, I arrived with the 335th in January 1968. I worked in maintenance and was a crewchief in the second platoon before being transferred to the 61st AHC. I joined the 3rd platoon, Starblazers and extended twice, serving my last six months as a Technical Inspector, mostly at LZ English. A rather sad story in the T.I. department, of the three guys, two of us were from Montana. We joked about the Montana boys at the 61st, but years later I discovered Pat perished on a flight and was never recovered. That hit close to home in many regards.

I discharged at Fort Lewis in March 1970. The Seattle Community College spring quarter started the next week and I enrolled, staying two years in Seattle. The Associate Aeronautical degree ended with my getting the A&P mechanic license. I flew at Galvin Flying Service located at Boeing Field and got the commercial, instrument rated, and certified flight instructor pilot licenses. My first wife and I met in Seattle, she was attending the U of W, but we moved to Missoula, Montana after two years in Washington.

I graduated from the U of M (Go Griz!) in Missoula with a Bachelor of Science degree in business administration—emphasis in accounting. The spring quarter of 1975, we still had a quarter system, I passed the Certified Public Accountant exam and got the Montana CPA license in 1977. I've been a California, Wyoming, and Montana CPA during my career, but the Montana license is the longest ago, now forty years!

I was interested in public service, so my first seven work years were with the US General Accounting Office (GAO—Congress auditors), and the Defense Contract Audit Agency (DCAA) in the Los Angeles region. The GAO work took me all over the western United States in a wide variety of assignments. I did lots of jobs with military aviation programs and airport facilities. The DCAA work evolved into a special detachment assignment dealing with all the highest classified programs managed out of Washington, DC, including CIA projects. I had a meeting with the director, William Casey, because he wanted to thank us for our work. He offered his help should any of us need him; that was a very special meeting and I left with his business card!

I audited spook projects like the GPS satellites, and stealth fighter and bomber projects, as a couple examples. My work became more rewarding when the projects became generally known and I could talk about them some. Spook projects of the late 1970s and early 1980s came to revolutionize navigation, joint service operations, precision armaments, ocean and weather studies, and many other aspects of our lives in the 21st century world.

Industry accounting in 1983 called and started working with government contractors until 1995. The companies ranged in size from small to Fortune 500 companies. I focused attention to uses of personal computers (PC) and software in industry applications. I believed the PC would revolutionize data handling for most uses in industry, government, and personal information management. I am proud of the many companies and people I've helped and trained in the exciting world of PC computing.

In 1995, my family moved from Santa Cruz to Stevensville. I had bought farm land in 1971 using my Army savings. I built a house on the property in 1981. I have been practicing accounting and tax work ever since. Sometimes, I have a crazy idea of becoming the oldest CPA in Montana, my certificate number is 1087! I especially enjoy the people I serve.

The grand Bitter Root Valley is a conservative community having a high percent of the veterans and otherwise there are lots of really good citizens. I enjoyed many hobbies over the years but have settled into RC airplanes I can fly off the back horse pasture. I enjoy assisting veterans to correct their records, and to get their VA services and disability benefits. Finally, I am proud to be a Vietnam veteran! I am blessed with many veteran friends.

My first marriage lasted thirty-one years, extending through most of our family's growing up years with three sons. Unfortunately, a woman from California (first wife) and a man from Montana (me) didn't mix well for retirement years and she returned to California in 2003. Our sons were entering college at that time, so our home was becoming an empty nest anyhow. I remarried in 2004 and that lasted nearly ten years. Now in 2017, I am looking to find a nice lady with whom to share my remaining years.

John, my oldest, is an MD at the Rochester Medical Center in New York. He is married, and they had a baby girl in November 2017. David, the middle son, has a Master's Degree in Software Engineering and has just finished working at Google in Seattle. He wants to become a university professor because he loves teaching. My youngest son, Mark, began ballet at age thirteen and now dances with the Norwegian National Ballet Company in Oslo. He is one of a few permanent career dancers there. He wants to return to the US and change his career to acting.

I found Christ as my savior while serving with the Starblazers in 1969, interesting how an experience like that led me into spiritual matters. We were losing enough men, so I felt the danger nearly every day. I surrendered my life to Christ as I witnessed terrible tragedies and after my best friend died in a crash. I've had ups and downs but overall my life has been greatly blessed. I am proud to have served with such distinguished units that made Army aviation history!

I served twelve years in youth baseball as a coach, manager, and league director. I have life membership in the American Legion, Disabled Veterans of American, and the Airplane Owners and Pilot Association. I am interested in being an amateur military historian from the Revolutionary War through the Vietnam War. Currently, I am the webmaster for the 61st AHC website at 61ahc.org. Our unit is proud of the Bong Son Lucky Stars Library and Learning Center that we built, then furnished a computer network. I love Christ, my country, family, community, and wish God's blessing upon all who might read these words. If anyone visits into Western Montana, please look me up online or in the phone directory.

WEBB, MONTY (TOM) WESLEY

My "Post Vietnam" life began with orders for Fort Rucker to work as a flight instructor where we taught the last two weeks of WOC training. Looking back this turned out to be one of the jobs I enjoyed most after leaving Vietnam. It was during this time that I got married and a year later would have a son.

After my military discharge, I took a short, good paying job as an Ironworker while sending out resumes to over 30 companies involved in helicopter operations. Jobs where hard to come by but finally I was able to hire on with Evergreen Helicopters in McMinnville, OR where we used Hiller 12Es, quite a comedown from the Huey. We worked in reforestation projects with the forest service in dangerous conditions for little pay and long hours.

I finally decided to move on and was hired by Topline Equipment Co, dealer for Clark forklifts and Michigan heavy equipment, in Portland, OR. I worked in a dual position as sales rep and company pilot where I flew a Hughes 500 and flew copilot in a Cessna 421 with the owner. I really enjoyed this position, but it would end in 1980 as the economy had taken a down turn and equipment sales became more and more difficult.

This became kind of a turning point in my life as I was offered a job in Phoenix, AZ working with United Rentals, construction equipment rental company as a rental rep. In 1984, I would go thru a divorce and relocate to Tucson, AZ as a store manager for the company. It was during this time I met my current wife of 32 years.

After three years as store manager, I decided to open my own business in Tucson, C & I Equipment Co, in sales of heavy equipment. In 1996, I came up with a design for a new 500-gallon water trailer with a polyethylene tank that would replace the outdated units with steel tanks. Water trailers were almost mandatory in Arizona and other southwestern states for dust control on job sites. Sales of this unit took off and I was able to drop out of heavy equipment and shift in to manufacturing. By 2007, we had become one of the leading manufacturers in the United States for water trailers. Unfortunately, in late 2007 our business burned to the ground because of a Linseed oil spill. It was at this point my wife and I decided to sell our business, retire, and move back to Washington.

We currently live in Port Angeles, WA about 60 miles west of Seattle, an ideal retirement area. We have a 27' travel trailer which gives us our freedom and a 19' boat which allows us to fish for Salmon and Halibut in the Straits of Juan De Fuca.

I would like to take this opportunity to say what an "Honor" it was for me to be part of the 335th Assault Helicopter Company. I take immense pride in talking about the "Cowboys" and the caliber of people I served with.

I'm proud to have served in the military and even more so every time I meet someone who says he wanted to join or wanted to be a helicopter pilot but for some reason just couldn't do it.

So many vets I've talked with are envious of how our company has stayed in contact, with a lot of credit going to Dom.

WEBER, WILLIAM FRANCIS

When I was in Phu Heip, I noticed that a soldier had his flack vest imprinted with the words 'Huntington Beach, CA is where it's happening' and I thought that it would be 'cool' if I were to go there.

In Vietnam, my ETS and DEROS were the same, so when I returned from overseas I was out of the service and eventually returned to Pittsburgh, PA, my hometown. So, I went back to my original job in St. Louis where I worked for McDonnell Douglas working on F-4s where I was originally drafted. They made good on giving me my job back as I was a returning veteran and were required to do so. When the opportunity came up to be transferred to an off-site base at Edwards A.F.B. in California, I grabbed it. I arrived there in March of 1969, one year after I got to Vietnam. I only spent seven months in Vietnam as my time of service was ended over there. However, my time at Edwards only lasted until August of 1969 when they 'Let me go'.

I complained that they had to keep me on for one year after my service ended, and even went to the federal building in L.A. to complain about it, but it was to no avail. The unemployment rate at this time was high in my area, so I worked a lot of different jobs to 'survive'. I laid carpet, pulled live turkeys out of a truck, lol, worked on the new freeway to Los Angeles, and worked at a plant making flares that we dropped from our ships in Vietnam to light up the skies. After a year, I finally got an excellent job at Lockheed aircraft here in Antelope Valley, CA building L-1011 commercial aircraft. This job lasted twelve years and I then transferred to the U-2 program where I ended up holding six different classifications and ended my time there launching these amazing planes into the air and recovering them. I had to retire after 39 yrs. there, 25 on the U-2, after my accident on my Harley left me with a broken leg and unable to run and retrieve the U-2 pogos (temporary wheels used to launch them). My career at Lockheed-Martin was kind of bittersweet as I realized that you're just a number there.

In 1983, I got married at the age of 37 and went on to have one daughter and one son that I'm very proud of. Unfortunately, my wife died in a car accident in 1989, which left me to raise the kids myself, with help from a good babysitter and my in-laws. My daughter was only two years old and son was only six months. Both ended up with college degrees and my son was an All-American baseball player at C.S.U.L.A. Over the years, I've played a variety of sports, including softball, snow skiing, and billiards as I grew up playing a lot of pool in Pennsylvania.

It's amazing the influence of only two years in the service had on my values of life and respect for our flag and what it stands for. At the time, I didn't realize it, but the respect and admiration has become a part of my life. Outside my house stands a 25-ft. flagpole with both the American flag and POW flag below it. We veterans are finally getting the appreciation that took so long to come.

WEBLEY, DANIEL CARL

Submitted by: Crystal Webley

In 1970, after being discharged from the military and the war in Vietnam, Daniel Webley returned to Seattle Washington. Within the first year after coming back home he lost both of his parents. His mother to cancer and his dad to a heart attack. He connected back with friends and started to work packing cod with his uncle and laying brick, building fireplaces and walls, and framing houses, etc.

He then married Christine Ellsworth who was a childhood friend. She had two young daughters at the time, Brenda and Tammy, whom Dan adopted. Dan and Christine also had two sons, Jim and Jacob. The Webley family then moved to Montana.

While in Montana, Dan begin logging and mining. Dan and Christine were divorced after ten years. Dan returned to the Seattle area and married Cathy Morris. She had a young son, DJ, whom Dan also adopted. They then moved to Nevada where they also helped to raise Dan's young niece, Tiffany.

Dan spent ten years working for the Barrick mine, driving heavy equipment. Kathy and Dan were married for 16 years and then divorced. Dan then moved back to Montana where he met his present wife, Crystal Huisentruit. Crystal is a professional musician and vocalist and had five children before meeting Dan. Dan became father to the five children and began his retirement in Troy, Montana. Dan's two sons and their wives and children also live in Troy, Montana and Dan's two daughters also live in Troy, Montana with their husbands and children.

The entire family loves to hunt and fish and gather as often as possible to have great family dinners and enjoy the fruits of labor from Crystal and Dan's gardens. Dan and Crystal have been married for 15 1/2 years. In 2007, they lost their daughter, Hannah, in a car accident and in 2017 they lost their granddaughter, Julie, to a car accident. Both girls had attended the same school and were friends.

Dan came back from Vietnam suffering from PTSD that he still has to this day, and nightmares. The war took quite an emotional and physical toll on him. But he is always described as a very kind, loving, humble, and jovial, humorous, and witty man. Currently, he is suffering from prostate cancer and problems with his lungs. He is enjoying his retirement watching his big screen TV sitting in his recliner and watching deer and wild turkeys cross in front of the picture window in the living room. He shares two acres, a creek, three gardens, and a little farmhouse cottage with his wife, Crystal.

Dan has always had a great admiration and love for young people, as you can see all the children and grandchildren that he has gathered in his life. Seven sons, three daughters, fourteen grandchildren and six great-grandchildren and many nieces and nephews, daughter-in-law's, and son-in-law's. Dan calls it his extended family! And in all their eyes, he is awesome! I would also like to add that Dan's oldest son is a Baptist minister and his son Jacob became a first time father at the age of 40! Way to go!

Thank you all for your service!

WILSON, DOUGLAS G.

I had two hitches in the Army. The first from March 1965 to March 1967, then again from April 1969 to April 1971. I also served in Vietnam with three different units. I was crewchief with the 502nd Avn, 114th AHC, 335th AHC, and 101st Airborne. Three gunship platoons Mavericks, Cobras, and Falcons and air ambulance with Eagle Dustoff.

I had a job before I went into the Army working for Cherry Rivet in the inspection department. I eventually went to the tool and die department and that was my occupation the rest of my working career. I changed factories a few times.

I had a midget race car and raced it 1971-73. I wasn't much of a driver, so I sold the car in 1974.

I took a four-month trip around the world in 1974. After the trip, I went to work for a small tool and die shop. I also rented a house in Rosarito Beach, Mexico and spent weekends and vacations there for five years.

When the tool and die shop moved to Spokane, WA, I went with them. I stayed from 1979-1985 then moved back to sunny California.

In 1992, I found a copy of old Army orders that had four names listed. I decided to look up the names to see if they were still around. Thereafter, I began my quest to locate more Falcons and Cowboys. This initial exercise helped in the first Falcon reunion in Phoenix, Arizona in 1993. I continue to locate former Vietnam Veterans for the Cowboys and Mavericks in my spare time.

I then worked for SPS Manufacturing and built a few street rods.

When SPS moved to Salt Lake, they didn't want to take their former employees. I was unemployed for about three years, so I built a roadster race car that I raced at El Mirage Dry Lake and Bonneville Salt Flats, Utah in 1996-98. I went 177 MPH in a car I built in my driveway.

Next, I went to work for Bristol Manufacturing until the recession, but after a few months I went to work for Fairchild. That was one of those seven day a week jobs. I don't think I lasted a year.

I was then able to go back to Bristol Manufacturing and stayed until I was 64 years old and retired in 2007. I am happily retired. My hobbies are my two 1932 Fords and one 1923 Ford roadster. My other hobby is IV Corps Vietnam History. I enjoy it and do some research on Vietnam every day. Even right now!

WOOD, JOHN LEONARD

(J. Leonard Wood)

September 1966 thru August 1967 in RVN (Republic of Vietnam). During that time, I was Platoon Leader, 2nd Airlift Platoon (Mustangs), 335th AHC (Assault Helicopter Company) known as the Cowboys. The 335th was attached to the 173rd Airborne Brigade, known as the Herd.

I am retired after 21 years from the Regular Army. I am now, at 80 years of age, ready to be on 'Easy Street', having worked for six companies since leaving the Army in 1979, all in the field of Aviation and Aviation Security.

The Army provided me opportunities to advance myself and I took them. I was a Field Artillery Officer, exposed to guns, howitzers, and missiles. The latter caused me to study several computer software languages such as FORTRAN, COBOL, and the like. I volunteered for specialty training as an Army Aviator. I graduated as a fixed wing and rotary wing qualified aviator. I went to combat in Vietnam. I was awarded three DFC (Distinguished Flying Cross). The Army sent me to Graduate School (Georgia Southern) where I was awarded an MBA (Master of Business Aviation). So, in 1979, I retired from the Army; the next day I went to work for the BDM Corporation, an aviation studies and analysis group. My military experience did influence my post-military work choice. In both the military and my later life, my military work set the stage for my work thereafter.

Summary of Civilian Work (1979 – 2017)

I knew I wanted to get Commercial Aviation credentials. To be at the top of the ladder among the pilot community, I needed the highest pilot certificate, an Airline Transport Pilot Certificate (ATP). Because I already had 2,000 hours as an airplane pilot and 2,200 pilot hours in helicopters, and because in planning for the ATP I flew some ATP training hours, the FAA (Federal Aviation Administration) awarded me an ATP. Now I could converse with Air Carrier Chief Pilots on their own turf, and with Air Carrier personnel at airports. In my spare time with the companies below, I flew jet charters (Learjet, Cessna Citations).

Calendar-wise I was employed in civilian aviation endeavors from the time I left the Army in July 1979 until December 2017. For reasons known only to the hiring agents, I was solicited for each of the following positions.

- The BDM Corporation
- Maryland Aviation Administration
- Ogden Corporation
- Condor Aviation Corporation
- Counter Technology Inc (CTI)
- Tyco/ADT Corporation

What drew me to these six companies? The facts are that each had a challenge for me in its mission. Each was more intense than the previous job and each successively paid a better salary. BDM wanted studies of aviation. Maryland Aviation, operator of BWI (Baltimore/Washington International Airport) wanted an Airport Manager. Ogden, a Fortune 500 firm, wanted an executive to provide aviation security to airports and air carriers. Condor was headed by me/founded by me, to provide security consulting to airports and air carriers. CTI wanted an executive to lead aviation security marketing. Tyco/ADT wanted a manager of aviation services — electronic services. Was there success? Resoundingly, YES. On a competitive basis, I was awarded these jobs.

Specific Details of Civilian Work (1979 — December 2017)

BDM Corporation, McLean, VA 1979-1981

As Aviation Operations Analyst, I managed and budgeted aviation analysis projects to include airport capacity and use studies for various aerospace companies.

Maryland Aviation Administration 1981-1989

Operations Manager, BWI 1981-1982

Administered, supervised, and coordinated the day-to-day operation, communications/information, fire-rescue, police, security, and maintenance activities of BWI, a Large Hub commercial airport.

Manager, Martin State (MTN) Airport, 1984-1985

Annual Operating Budget: $ 3M capital; $ 2M O&M

Directed this large general aviation (GA) reliever airport, one of Maryland's busiest (170,000 air operations/year). Included management of airport-owned FBO (1 million-gallon annual sales). Responsible for Mutual Aid Agreements. Used Federal grants to accomplish runway, taxiway, and ramp design, and construction.

Manager, BWI Airport, 1982-1984

Annual Operating Budget: $ 25M capital; $ 27M O&M.

A Large-Hub airport having five million enplanements; 300,000 air operations: 31 air carriers; 215 tenant organizations; also served domestic and international charter carriers (265,000 deplaned; 200,000 enplaned). Ensured that BWI's airport operation and its security continually complied with FARs (Federal Aviation Regulations), especially Parts 139, 107, 108, and 77 daily, without error. Responsible as well for all landside operations, and Mutual Aid Agreements. Created specs for BWI's security access control system. Developed, coordinated and executed BWI's ASP (Airport Security Program).

Associate Administrator—Operations, 1985-1989

Annual Operating Budget: $30M capital; $30M O&M.

As head of a 263-person airport operation element, planned, organized, staffed, directed, and controlled Baltimore/Washington International (BWI) and MTN airports as to operations, construction and maintenance at this Large Hub commercial airport. This included the licensing, annual certification, and operational oversight of 37 other public-use airports in Maryland; plus, the registration of 60+ private-use airports. Interacted with 24 county governments due to my operational oversight of the public-use airports in Maryland. Left State service to form Sander, Wood & Associates.

Ogden Corporation, Arlington, VA 1989-1993

Ogden, a Fortune 500 firm, bought Sander, Wood & Associates, Inc., a security consulting firm which was founded in 1989 by Messrs. Sander and Wood who had been Maryland Aviation Administration executives. Its charter is to provide security consulting to air carriers (FAR 108) and airports (FAR 107). Work included airport security system and communication center planning and design, and construction/installation oversight, of airport police facilities, security, and computerized/automated systems and ASP development and coordination. Clients included: Chicago O'Hare, Detroit Metropolitan, Memphis International, Chicago Midway, Dane County (Madison, Wisconsin), Washington Dulles, T. F. Green (Providence, Rhode Island), Albany County (Albany, New York), Palm Springs, Peoria, Lansing, Flint, Grand Rapids, Saginaw, Kalamazoo/Battle Creek, Detroit City, and Traverse City. With Ogden, the client base was expanded by Wood and Sander to include: Milwaukee, Lexington, Minneapolis, Wichita, and US Virgin Islands (St Thomas; St Croix) airports. Both Messrs. Sander and Wood departed Ogden in 1993 when Ogden chose to discontinue its consulting practice.

Condor Aviation Corporation, Timonium, MD July 2006—2012

I formed this company to provide Airport Security Consulting services as a subcontractor to firms engaged in airport industry projects. Such firms included: Architectural/Engineering; Security Engineering; Airport Planning firms. My services included: application of federal regulations and guidelines to airport security projects, at Commercial as well as General Aviation airports; composition of Airport Security Program (ASP) documents; commentary on proposed specifications/text and plans/drawings; airport security project management; Third-Party review of planned airport security projects, as well as Third-Party oversight of projects underway. Also provided Airport Operation Consulting. Airport security projects by me in 2006-2012 included: Fort Lauderdale (FLL) action standards manual; St Thomas (STT) and St Croix (STX) airports new ASP; a year-long study of International General Aviation Security for the TSA; employee manuals for the PANY&NJ; and Midfield Satellite Concourse project at LAX (utility plant security design).

Counter Technology, Inc, Bethesda. MD 1994-2001

As Director, Facility Planning-Management-Operations, I provided airport and air carrier security consulting. Clients included Austin-Bergstrom, Boston, Chicago-O'Hare, Chicago Midway, Houston Intercontinental, Jackson MS, MidAmerica St Louis (BLV), Phoenix, Roanoke, Ronald Reagan Washington National, and Washington Dulles airports. Work included airport security system, communication center planning and design, and construction/installation oversight, of airport police facilities, security and computerized/automated systems; also, ASP development and coordination. For air carriers, work included ramp operation planning, back-office security planning and design, and the air carrier-to-airport interface. 'Pro bono' work included participation during 1994-1999 in the FAA/ACC Airport

Terminal Planning Process work group; and speaking at national-level aviation conferences.

Tyco/ADT, Alexandria, VA 2001-July 2006

As Manager, Aviation Services — ADT nationwide, I marketed Tyco/ADT electronic security systems equipment, installation and construction services, and maintenance services, to airports and air carriers. My ADT work required me to know TSA security regulations and the technical aspects of security systems to respond to airport RFPs for facility security systems (access control, CCTV, intercom, video ID) equipment, installation and maintenance. Participated in Airport Access Control projects at GA airports (Martin State, Baltimore) and Commercial airports (Wilkes-Barre/Scranton).

Condor – Timonium, MD 2006-December 2017

Accomplished pro bono work for RTCA to constantly update and improve, as a permanent committee of the RTCA, the RTCA Standards for "Airport Security Access Control," a national-level work.

Did pro bono work for the TRB (Transportation Research Board — Committee AV090). Defined roles and missions for research projects affecting aviation.

Closing

This document tells the lifetime work history of J. Leonard Wood.

In closing, several post-script matters must be told. First, I am a member of the professional organization, "American Association of

Airport Executives (AAAE)," 1981 to present. To be a member, one must have served as an Airport executive. AAAE meets annually. Second, I am a member of a social organization, "The Cowboys," who served 1965-1972 with the 335th AHC in the Republic of Vietnam in any capacity. "The Cowboys" meet every two years, most recently in 2017.

I am a resident of Maryland.

I will continue my interest in aviation and my membership in "The Cowboys"

GLOSSARY

A&P: Airframe and Powerplant

AAF: Army Air Field

AARP: Association of American Retired Persons

AAS: Associate of Applied Science

ABC: American Broadcasting Company

ABN: Airborne

ACC: Air Craft Commander

ADT: American District Telegraph

AFB: Air Force Base

AFL-CIO: The American Federation of Labor and Congress of Industrial Organizations

AFROTC: Air Force Reserve Officer Training Corps

AGL: In aviation, atmospheric sciences and broadcasting, a height above ground level (AGL)

AHC: Assault Helicopter Company

AHP: Army Heliport

AIT: Advanced Individual Training

ALCAN: Alaska-Canadian Highway

AMMEDS: U.S. Army Medical Department Center and School

AMVETS: American Veterans

AOPA: Aircraft Owners & Pilots Association

APO: Army Post Office

ARMISH-MAAG: Mission assigned to the Iranian Ministry of War, known as ARMISH, for the purpose of training the Iranian army. With the inception of the U.S. military assistance program an

American advisory group, MAAG, was formed in 1950 to administer the flow of arms into Iran.

ARVN: Army of the Republic of Vietnam

AV. Or AVN: Aviation

AVG: American Volunteer Group

AVR: Aortic valve replacement

AVSCOM: United States Army Aviation Systems Command

AWOL: Absent Without Official Leave

BCTP: Battle Command Training Program

Bde: Brigade

Bn: Battalion

BNCOC: Basic Non-Commissioned Officer Course

BRAC: Base Realignment And Closure

BSEE: Bachelor of Science in Electrical Engineering

BWI: Baltimore Washington International

CA: Combat Assault

CAD: Computer Aided Design

CAV: Cavalry

CCTV: Closed Circuit TV

CDSO: Collateral Duty Safety Officer

CEO: Chief Executive Officer

CFI: Certificated (or Certified) Flight Instructor

CGSC: Command and General Staff College

CIA: Central Intelligence Agency

CID: The United States Army Criminal Investigation Command (USACIDC, usually abbreviated as just CID)

CO: Commanding Officer

CONUS: Continental United States or Contiguous United States

CPA: Certified Public Accountant

CPT: Captain

CTI: Counter Technology Inc.

CWO: Chief Warrant Officer

DA: Department of Army

DAV: Disabled American Veterans

DBA: Doing Business As

DEA: Drug Enforcement Administration

DEROS: Date Eligible for Return From Overseas

DFC: Distinguished Flying Cross

DFCS: Distinguished Flying Crosses - plural

DFW: Dallas/Fort Worth

DIV or Div.: Division

DJJ: Department of Juvenile Justice

DMZ: Demilitarized Zone

DoD: Department of Defense

DOE: Department of Energy

EAA: Experimental Aircraft Association

EEOC: Equal Employment Opportunity Commission

EIC: Excellence in Competition

EKG: Electrocardiogram

EM: Enlisted Military

EMS: Emergency Medical Services

EMT: Emergency Medical Technician

EMU's: Experimental Military Unit (Austrian Army)

ENT: Ear, Nose, Throat

ETS: Expiration of Term of Service

EVS: Environmental Studies

FAA: Federal Aviation Administration

FBO: A fixed-base operator (FBO) is an organization granted the right by an airport to operate at the airport and provide aeronautical services such as fueling, hangaring, etc.

FNG: Freaking New Guy

FORSCOM: United States Army Forces Command

FRG: Federal Republic of Germany

GAO: Government Accountability Office

GDP: Ground Defense Position

GED: General Equivalency Development or General Equivalency Diploma

GIS: Geographical Information System

HAAF: Hunter Army Airfield

HIFC: Helicopter Instrument Flying Course

HMFIC: Head Man Fully In Charge (there is another definition - not proper to include in this book)

HQ: Headquarters

HQIFFORCEV: Headquarters 1st Field Force Vietnam

IFE: Instrument Flight Examiner

IFR: Instrument Flight Rules

IG: Inspector General

ILS: Instrument Landing System

INS: Immigration and Naturalization Service

IP: Instructor Pilot

IRR: Individual Ready Reserve

IRS: Internal Revenue Service

IRT: Instrument Refresher Training

JROTC: Junior Reserve Officer Training Corps

JTF: Joint Task Forces

KATL: Hartsfield - Jackson Atlanta International Airport

KMAG: Korean Military Advisory Group

KMEM: Memphis Intl Airport

KPDK: Dekalb/Peachtree. Located in Atlanta, GA

LRRP: A long-range reconnaissance patrol, or LRRP (pronounced "lurp")

LTC: Lieutenant Colonel

LZ: Landing Zone

MACV: Military Assistance Command, Vietnam

MAG: Mecaer Aviation Group

Maj: Major

MASH: Mobile Army Surgical Hospital

MIA: Missing In Action

MOAA: Military Officers Association of America

MOI: Maintenance Operating Instruction

MOS: Military Occupational Specialty

MP: Military Police

MPA: Master of Public Administration

MSP: Maryland State Police

MSW: Master of Social Work

NATO: North Atlantic Treaty Organization

NCGA: Northern California Golf Association

NCNG: North Carolina National Guard

NCO: A non-commissioned officer or noncommissioned officer (NCO, colloquially non-com or noncom)

NCOIC: Non-Commissioned Officer in Charge

NG: National Guard

NK: North Korean

NMC: Naval Medical Center

NRA: National Rifle Association

NSA: National Security Agency

NVA: North Vietnamese Army

OAK: Oakland International Airport

OCS: Officer Candidate School

OER: Officer Evaluation Report

OJT: On the Job Training

OLC: Oak Leaf Cluster

OPS: Operations

PBS: Public Broadcasting Service

PCS: Permanent Change of Station

PFC: Private First Class

PGR: Patriot Guard Riders

PIC: Pilot In Command

POW: Prisoner of War

PTSD: Posttraumatic Stress Disorder

PX: Post Exchange

R&R: Military slang for rest and recuperation or rest and relaxation or rest and recreation

RIF: Reduction in Force

RLO: Regional Liaison Officer

ROK: Republic Of Korea

ROTC: Reserve Officer Training Corps

RPG: Rocket-propelled Grenade

RSO: Reconnaissance Surveillance Officer

RTCA: Radio Technical Commission for Aeronautics

RVN: Republic of Vietnam

SCARNG: South Carolina Army National Guard

SFC: Sergeant First Class

SGM: Sergeant Major

SGT: Sergeant

SIP: Standardization Instructor Pilot

SIU: Southern Illinois University

SOPS: Standard Operating Procedures - plural

SPC: Specialist

SSG: Staff Sergeant

TAC: Short for "tactical", it is an acronym for: T – Teach   A – Advise C - Counsel

TC: Training Coordinator

TDY: Temporary Duty

TRADOC: Training & Doctrine Command

Trans Det.: Transportation Detachment

TROSCOM: Troop Support Command

TSA: Transportation Security Administration

TWD: Tak Won Do

UDT: Underwater Demolition Team

UHF: Ultra-High Frequency
UMC: United Methodist Church
USAA: United Services Automobile Association
USAPHC: United States Army Primary Helicopter Center
USAR: United States Army Reserve
USARPAC: United States Army, Pacific
USC: University of Southern California
USDA: United States Department of Agriculture
USFS: United States Forest Service
USGS: United States Geological Survey
USMC: United States Marine Corps
USS: United States Ship
VA: Veterans Administration
VAMC: Veterans Administration Medical Center
VC: Viet Cong
VFR: Visual Flight Rules
VFW: Veterans of Foreign Wars
VHPA: Vietnam Helicopter Pilot Association
VIPs: Very Important Person
VNAF: Vietnamese Air Force
WO: Warrant Officer
WOC: Warrant Officer Candidate
XO: Executive Officer

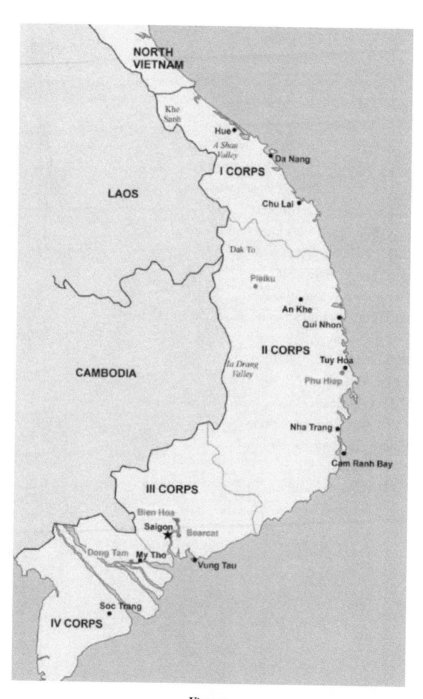

Vietnam

THE 335TH ASSAULT HELICOPTOR COMPANY

Top Row: James Gann, Don Sester, Darrell Condry, Tom Schiers, John Dibble, Dave Fraser. Craig Kramer, Rick Dorer, Tom Gould, Milton Smith, Ed Grabowski, Don Dowdy, Dennis Stogner, Don Snyder, Neil Leonard, & Thomas Bothur

Center Row: Jack Hunnicutt, Mark Schimpf, George Mayl, Howard Stiles, Bob La Brie, Julian Perez, Dom Fino, Andy Pregillana, Frank Wannomae, Neil Stribling, Gary Peyton, George Murray, & Vance Gammons

Front Row: Roly Linstad, Bob Murphy, Bob McClellan, James O'Garra, Robert Smith, Jim Stein, Lee LaGrelius, Paul Millen, Andy Hooker, Gale Smith, Jim Quattlebaum, Dennis DuPuis, Ron Corb, Ralph Luffman, & Norm Polacke

CPSIA information can be obtained
at www.ICGtesting.com
Printed in the USA
BVHW03s2232190618
519329BV00046B/674/P